Lecture Notes in Mathematics

Edited by A. Dold and B. Eckmann

1160

Jan-Cees van der Meer

The Hamiltonian Hopf Bifurcation

Springer-Verlag
Berlin Heidelberg New York Tokyo

Author

Jan-Cees van der Meer
Centre for Mathematics and Computer Science (CWI)
Konislaan 413, 1098 SJ Amsterdam
The Netherlands

Mathematics Subject Classification (1980): 58F05, 58F14, 58F22, 58F30, 70F07

ISBN 3-540-16037-X Springer-Verlag Berlin Heidelberg New York Tokyo
ISBN 0-387-16037-X Springer-Verlag New York Heidelberg Berlin Tokyo

Library of Congress Cataloging in Publication Data. Meer, Jan-Cees van der, 1955-. The Hamiltonian Hopf bifurcation. (Lecture notes in mathematics; 1160) Bibliography: p. Includes index. 1. Hamiltonian mechanics. 2. Bifurcation theory. I. Title. II. Series: Lecture notes in mathematics (Springer-Verlag); 1160.
QA3.L28 no. 1160 [QA614.83] 510 s [514'.74] 85-27646
ISBN 0-387-16037-X (U.S.)

Printing and binding: Beltz Offsetdruck, Hemsbach / Bergstr.
2146/3140-543210

Preface

The research laid down in these notes began several years ago with some questions about a particular bifurcation of periodic solutions in the restricted problem of three bodies at the equilibrium L_4 . This particular bifurcation takes place when, for the linearized system, the equilibrium L_4 changes from stable to unstable. This kind of bifurcation is called a Hamiltonian Hopf bifurcation.

During the research it became apparent that new methods had to be developed and that existing methods had to be reformulated in order to deal with the specific nature of the problem. The development of these methods together with their application to the Hamiltonian Hopf bifurcation is the main topic of these notes. As a result a complete description is obtained of the bifurcation of periodic solutions for the generic case of the Hamiltonian Hopf bifurcation.

This research was carried out at the Mathematical Institute of the State University of Utrecht. I am very grateful to Prof. Hans Duistermaat and Dr. Richard Cushman for their guidance and advice during the years I worked on this subject. I also thank Richard Cushman for his careful reading of the earlier drafts of the manuscript. Thanks are also due to Prof. D. Siersma of the University of Utrecht for the discussions we had on chapters 3 and 4 , and to Prof. F. Takens of the University of Groningen for his remarks concerning the final manuscript. Finally, I would like to thank Drs. H. van der Meer for his assistence in plotting fig. 4.1 - 4. 14, and Jacqueline Vermeij and Jeannette Guilliamse for their excellent typing of the manuscript.

Jan-Cees van der Meer

June 1985

CONTENTS

Introduction

In this monograph the main topic is the study of periodic solutions of a family of Hamiltonian systems of two degrees of freedom near an equilibrium as the family passes through resonance. We concentrate on the Hamiltonian Hopf bifurcation, that is, the passage through nonsemisimple 1 : -1 resonance. The nonsemisimple 1 : -1 resonance distinguishes itself from the other resonances in two ways: first, at the resonance the linearized system is nonsemisimple having two equal pairs of purely imaginary eigenvalues; second, when passing through resonance, the equilibrium point changes from elliptic to hyperbolic type. Although we have concentrated on a specific example, many parts of the theory developed here have much wider applications, especially to other two degree of freedom resonances.

The approach we take can be divided into four main parts: (1) formal normal form theory; (2) equivariant theory of stability of maps applied to energy-momentum maps to derive standard systems; (3) geometric treatment of the standard system; and (4) Moser-Weinstein reduction to extend the results to nonintegrable systems.

The general formal normal form theory for Hamiltonian systems is treated first (chapter 2). Here we focus on the normalization of the Hamiltonian function. We do not restrict ourselves to systems of two degrees of freedom. The theory is illustrated by the classical examples of Hamiltonian systems with purely imaginary eigenvalues.

If we consider the Hamiltonian $H = H_2 + H_3 + \dots$ of a Hamiltonian system of two degrees of freedom then we may normalize H with respect to H_2 up to arbitrary order. Truncation then gives an integrable system provided that the semisimple part S of the homogeneous quadratic term H_2 is nonzero. If we consider the one parameter group $\mathcal{S}$ generated by the flow of the Hamiltonian vector field X_S corresponding to the integral S then the truncated normalized Hamiltonian $\tilde{H}$ is S-invariant. For the system corresponding to $\tilde{H}$ we consider the S-invariant energy-momentum map $\tilde{H} \times S$. To this energy-momentum map we apply the equivariant theory of stability of maps. For the case of the nonsemisimple 1 : -1 resonance we show that this energy-momentum mapping is finitely determined. The integrable system corresponding to the determining jet is called a standard system for the resonance (chapter 3).

In applying the theory of stability of maps, we drop the condition that the transformations used be symplectic. However much of the qualitative behaviour of the standard system can be translated back to the original system, especially the behaviour of periodic solutions. Using the theory of unfoldings we are able to study the behaviour of families of periodic solutions during the passage through resonance. The unfolded standard system for the nonsemisimple 1 : -1 resonance is studied in detail (chapter 4).

Finally we use some ideas of Weinstein and Moser to show how the periodic solutions of an arbitrary family of nonintegrable systems passing through resonance correspond to the periodic solutions of a family of integrable systems to which we may apply the preceding theory. This reduction from a nonintegrable to an integrable system in the search for periodic solutions is called the Moser-Weinstein reduction(chapter 5). The final result is a complete description of the behaviour of periodic solutions of short period in the generic case of the Hamiltonian Hopf bifurcation. Such a bifurcation appears in the restricted problem of three bodies at an equilateral equilibrium when the mass parameter passes through the critical value of Routh. It is this problem in the restricted problem of three bodies which inspired this study . Although combining all known results gave a fairly good description of the behaviour of periodic solutions (partially based on numerical results), a complete treatment and proof was nowhere to be found.

Because of the special properties of the nonsemisimple 1 : -1 resonance a new approach had to be followed. Many of the methods which had been succesfully used for the other resonances did not apply in this case. For the methods developed the nonsemisimple 1 : -1 resonance is the simplest example in the hierarchy of resonances, especially if one considers the computation of co-dimension and the geometric treatment of the standard system. The application of the normal form theory is a bit more complicated but the resulting normal form takes a simpler form than in the other resonances.

The chapters are organized as follows. In the first chapter preliminaries from the theory of Hamiltonian systems are treated. In the second chapter one finds the theory of Hamiltonian normal forms. In the third chapter the equivariant theory of stability of maps is applied to energy-momentum maps invariant with respect to a symplectic S^1-action. The fourth chapter deals with the geometry of the standard integrable system for the Hamiltonian Hopf bifurcation. Chapters 2,3 and 4 can be read independently. In chapter five the Moser-Weinstein reduction is applied to the Hamiltonian Hopf bifurcation. Together with the results of chapters 2,3 and 4 this leads to the main theorem (ch. 5, sect. 3). In chapter six we show how the theory applies to the restricted problem of three bodies. We conclude with a discussion of the known results concerning the nonsemisimple 1 : -1 resonance.

Chapter I

Preliminaries

0. Introduction

In this first chapter we will give a review of some facts from Hamiltonian mechanics which are fundamental to what follows. Emphasis is laid upon the relation between the symplectic geometric and the Lie algebraic features induced by the presence of the symplectic form. Also linear Hamiltonian systems are treated because they are basic for many features of and techniques used for nonlinear systems.

Most definitions and theorems are stated without proof. For the proofs and a more detailed treatment of the theory we refer to the text-books of Arnold [1978] and Abraham and Marsden [1978].

1. Hamiltonian systems

Consider the following system of ordinary first order differential equations on $\mathbb{R}^{2n}$

$$\frac{dq_i}{dt} = \frac{\partial H(q,p)}{\partial p_i} \tag{1.1}$$
$$\frac{dp_i}{dt} = -\frac{\partial H(q,p)}{\partial q_i} \quad ; \quad i = 1,\ldots,n.$$

where $H(q,p)$ is some real valued function on $\mathbb{R}^{2n}$, at least once differentiable. We call (1.1) a *Hamiltonian system of differential equations*. The function H in (1.1) is called a *Hamiltonian function*. The right hand side of (1.1) can be written as

$$X_H(q,p) = J.dH(q,p) \tag{1.2}$$

with

(1.3) $$J = \begin{pmatrix} 0 & I_n \\ -I_n & 0 \end{pmatrix}$$

where I_n is the $n \times n$ identity matrix. We call X_H the *Hamiltonian vector field associated to the Hamiltonian* H.

The above is the classical definition of Hamiltonian systems on $\mathbb{R}^{2n}$. This can also be obtained from the following more general differential geometric approach defining a Hamiltonian system on a manifold M.

Let ω be a two-form on M. We say that ω is *nondegenerate* if ω is a nondegenerate bilinear form on the tangent space of M at m for each $m \in M$. If there is a nondegenerate two form on M then M has even dimension. Furthermore we say that a two-form ω is *closed* if $d\omega = 0$ where d is the exterior derivative.

1.4. DEFINITION. A *symplectic form* ω on a manifold M is a nondegenerate closed two-form ω on M. A *symplectic manifold* (M,ω) is a manifold M together with a symplectic form ω on M.

1.5. DEFINITION. Let (M,ω) be a symplectic manifold and $H : M \to \mathbb{R}$ a C^k-function, $k \geq 1$. The vector field X_H determined by $\omega(X_H,Y) = dH.Y$ is called the *Hamiltonian vector field with Hamiltonian function* H. We call (M,ω,H) a *Hamiltonian system. We will suppose* H *to be* C^∞ *in the following.*

The following theorem shows that locally definition 1.5. is equivalent to the classical one.

1.6. THEOREM (Darboux). Let (M,ω) be a symplectic manifold then there is a chart (U,φ) at $m \in M$ such that $\varphi(m) = 0$ and with $\varphi(u) = (x_1,\ldots,x_n,y_1,\ldots,y_n)$ we have $\omega|U = \sum_{i=1}^{n} dx_i \wedge dy_i$.

The charts (U,φ) are called *symplectic charts* the coordinates x_i,y_i are called *symplectic* or *canonical coordinates*. Notice that if x_i,y_i are canonical coordinates then $X_H(x_i,y_i) = (\frac{\partial H}{\partial y_i}, -\frac{\partial H}{\partial x_i}) = J.dH$ with J given by (1.3).

We now define the notion of a flow of a Hamiltonian vector field together with some related notions. The flow in fact gives us the simultaneous motion in time of all points of M along the trajectories of the vector field.

1.7. DEFINITION. Let $\gamma(t)$ be a curve in $\mathbb{R}^{2n}$. We say that γ is an *integral curve for* X_H if $\frac{d\gamma}{dt} = X_H(\gamma)$, that is, if Hamilton's equations hold. Let (M,ω,H) be a Hamiltonian system. The map $\varphi : \mathbb{R} \times \mathbb{R}^{2n} \to \mathbb{R} \times \mathbb{R}^{2n}$ such that $\varphi_m : t \mapsto \varphi(t,m)$ is an integral curve at m for each $m \in M$ is called the *flow of* X_H. The curve $t \mapsto \varphi(t,m)$ is called the *maximal integral curve of* X_H *at* m or *the orbit of* X_H *through* m. The picture of M decomposed into orbits is known as the *phase portrait of* X_H.

Notice that the set $\{\varphi_t | t \in \mathbb{R}\}$ is a one-parameter group of diffeomorphisms of M, if every maximal integral curve is defined for all $\mathbb{R}$.

1.8. DEFINITION. A C^∞-map $\psi: (M,\omega) \to (M,\omega)$ is *symplectic* or *canonical* if $\psi^*\omega = \omega$.

Here $\psi^*\omega$ is the *pull-back* of ω under ψ defined by $\psi^*\omega(m)(e_1,e_2,\ldots,e_{2n}) = \omega(\psi(m))(d\psi(m)e_1,\ldots,d\psi(m)e_{2n})$. For $F \in C^\infty(M,\mathbb{R})$ $\psi^*F = F\circ\psi$. We have $\psi^*X_H = X_{\psi^*H} = X_{H\circ\psi}$, if ψ is symplectic.

It is clear that φ_t, $t \in \mathbb{R}$, defined by the flow φ of the Hamiltonian vectorfield X_H is a symplectic diffeomorphism. Note that $H(\gamma(t))$ is constant in t along integral curves $\gamma(t)$ of X_H. This corresponds to conservation of energy.

The following definitions show how the presence of a symplectic form on M induces a Lie algebra structure on $C^\infty(M,\mathbb{R})$ in a natural way.

1.9. DEFINITION. Let (M,ω) be a symplectic manifold and let $F,G \in C^\infty(M,\mathbb{R})$. The *Poisson bracket of* F *and* G is

$$\{G,F\} = \omega(X_F,X_G)$$

In canonical coordinates

$$\{G,F\} = \sum_{i=1}^{n} \left(\frac{\partial F}{\partial x_i}\frac{\partial G}{\partial y_i} - \frac{\partial F}{\partial y_i}\frac{\partial G}{\partial x_i}\right)$$

Notice that we have

$$\{G,F\} = dF.X_G$$

It follows directly that F is constant along orbits of X_G (or G constant along orbits of X_F) if and only if $\{F,G\} = 0$. $\{F,F\} = 0$ corresponds to conservation of energy for the system (M,ω,F).

1.10. DEFINITION. $F \in C^\infty(M,\mathbb{R})$ is an *integral for the system* (M,ω,H) if $\{H,F\} = 0$.

The notion of Poisson bracket allows us to consider the real vector space $C^\infty(M,\mathbb{R})$ as a Lie algebra.

1.11. DEFINITION. A *Lie algebra* is a vector space V with a bilinear operation [,] satisfying:
$[X,X] = 0$ for all $X \in V$ and
$[X,[Y,Z]] + [Y,[Z,X]] + [Z,[X,Y]] = 0$ (the "Jacobi identity")
for all $X,Y,Z \in V$.

It is now easily checked that $C^\infty(M,\mathbb{R})$ considered as a real vector space together with the Poisson bracket is a Lie algebra. Notice that the fact that ω is a closed two-form is essential in order to establish the Jacobi identity.

If ψ is symplectic then $\psi^*\{F,G\} = \{\psi^*F,\psi^*G\}$ for all $F,G \in C^\infty(M,\mathbb{R})$, that is, ψ^* is a Lie algebra isomomorphism. In fact the converse also holds.

On the space of Hamiltonian vector fields one has the usual Lie bracket of vector fields making this space into a Lie algebra. We have

(1.12) $$[X_F,X_G] = X_{\{F,G\}}$$

We call $[X_F,X_G]$ the *Lie bracket of* X_F *and* X_G. The Hamiltonian vector fields with Lie bracket form a Lie subalgebra of the Lie algebra of all vector fields. Notice that this Lie subalgebra is homomorphic to the Lie algebra $C^\infty(M,\mathbb{R})$ with Poisson bracket.

Returning to the Lie algebra $C^\infty(M,\mathbb{R})$ we may define for each $F \in C^\infty(M,\mathbb{R})$ the map $\mathrm{ad}(F) : C^\infty(M,\mathbb{R}) \to C^\infty(M,\mathbb{R})$ by $\mathrm{ad}(F)(G) = \{F,G\}$. The map $\mathrm{ad} : F \to \mathrm{ad}(F)$ is called the adjoint representation of $C^\infty(M,\mathbb{R})$. Because of the Jacobi identity $\mathrm{ad}(F)\{G,H\} = \{\mathrm{ad}(F)(G), H\} + \{G,\mathrm{ad}(F)(H)\}$ for each $G,H \in C^\infty(M,\mathbb{R})$, $\mathrm{ad}(F)$ is an inner derivation of $C^\infty(M,\mathbb{R})$ for each $F \in C^\infty(M,\mathbb{R})$.

1.13. REMARK. In the special case when M is a vector space we speak of a *symplectic vector space*. As before we may introduce the notions of Hamiltonian function, Hamiltonian vector field and Poisson bracket. Here we have global coordinates so these notions can be defined in terms of coordinates.

1.14. REMARK. Notice that our definition of Poisson bracket (definition 1.9.) differs from the one in Abraham and Marsden [1978] by a minus sign. This is done in order to obtain formula (1.12.) which gives rise to the Lie algebra homomorphism between Hamiltonian functions and Hamiltonian vector fields. Our definition agrees with Arnold [1978] if one takes into account that his standard symplectic form differs from ours by a minus sign.

According to Dugas [1950] our conventions agree with those of Poisson. Studying other literature it becomes clear that historically both conventions for Poisson bracket have been used.

2. Symmetry, integrability and reduction

In this section we will restrict ourselves to $\mathbb{R}^{2n}$ with coordinates $(x,y) = (x_1,\ldots,x_n,y_1,\ldots,y_n)$ and standard symplectic form $\omega = \sum_{i=1}^{n} dx_i \wedge dy_i$. Then $(\mathbb{R}^{2n},\omega)$ is a symplectic vector space as well as a symplectic manifold and $C^\infty(\mathbb{R}^{2n},\mathbb{R})$ with Poisson bracket as given by definition 1.9. is a Lie algebra.

In the following proposition some statements about Lie series are collected. The proofs are straight forward and left to the reader as an exercise. We define the *Lie series* $\exp \operatorname{ad}(H) = \sum_{n=0}^{\infty} \frac{1}{n!} \operatorname{ad}^n(H)$.

1.15. PROPOSITION. (i) $\operatorname{ad}(H)(x,y) = X_H(x,y)$ where $\operatorname{ad}(H)(x,y)$ is defined as $(\operatorname{ad}(H)x_1,\ldots,\operatorname{ad}(H)y_n)$.

(ii) $\exp(t\,\operatorname{ad}(H))(x,y)$ is the flow of X_H

(iii) $(F\circ\exp(\operatorname{ad}(H)))(x,y) = \exp(\operatorname{ad}(H))(F(x,y))$

(iv) $\exp(\operatorname{ad}(H))$ and $\exp(\operatorname{ad}(F))$ commute if and only if $\{H,F\}$ is constant.

In the last statement of the above proposition one might replace the condition $\{H,F\}$ is constant by $[X_H,X_F] = 0$ where $[\,,\,]$ is the Lie bracket given by (1.12). Proposition 1.15.(iv) is then equivalent to the statement that two Hamiltonian vector fields commute in the Lie algebra of vector fields if and only if their flows commute.

Now recall that the space of all maps $\operatorname{ad}(F)$, $F \in C^\infty(\mathbb{R}^{2n},\mathbb{R})$ is a Lie algebra with bracket $[\operatorname{ad}(F),\operatorname{ad}(G)] = \operatorname{ad}(\{F,G\})$. Therefore we have a group A generated by the $\exp(\mathrm{ad}(F))$, $F \in C^\infty(\mathbb{R}^{2n},\mathbb{R})$. Each one-parameter group $\exp(t\,\mathrm{ad}(F))$, $t \in \mathbb{R}$ forms a one-parameter subgroup of A. On the symplectic space $\mathbb{R}^{2n}$ each one-parameter group of diffeomorphisms is the flow of a Hamiltonian vector field. Thus we have found all one-parameter subgroups of A because each generator of A is a symplectic diffeomorphism which is the time one flow of a Hamiltonian vector field by prop. 1.15.(ii).

On $(\mathbb{R}^{2n}, \omega)$ let $\Phi : G \times \mathbb{R}^{2n} \to \mathbb{R}^{2n}$ be a *symplectic action* of the Lie group G on $\mathbb{R}^{2n}$, that is, for each $\varphi \in G$ the map $\Phi_\varphi : \mathbb{R}^{2n} \to \mathbb{R}^{2n} : x \mapsto \Phi(\varphi x)$ is symplectic. In a natural way the action Φ induces an action $\Psi : G \times C^\infty(\mathbb{R}^{2n},\mathbb{R}) \to C^\infty(\mathbb{R}^{2n},\mathbb{R}) : (\varphi,H) \mapsto H\circ\Phi_\varphi$ of G on $C^\infty(\mathbb{R}^{2n},\mathbb{R})$. In the following we will write $\varphi.H$ for $\Psi(\varphi,H)$.

1.16. <u>DEFINITION</u>. A Lie group G acting symplectically on $\mathbb{R}^{2n}$ is a *symmetry group for the system* $(\mathbb{R}^{2n},\omega,H)$ if $\varphi.H = H$ for all $\varphi \in G$.

Proposition 1.15. gives

1.17. <u>PROPOSITION</u>. If F is an integral for the system $(\mathbb{R}^{2n},\omega,H)$ then the one-parameter group $\exp(t\,\mathrm{ad}(F))$, $t \in \mathbb{R}$, given by the flow of F, is a symmetry group for $(\mathbb{R}^{2n},\omega,H)$.

The converse of proposition 1.17. also holds in the sense that each symmetry group of a Hamiltonian system gives rise to an integral. To make this precise we first introduce the notion of momentum mapping.

1.18. <u>DEFINITION</u>. On $(\mathbb{R}^{2n},\omega)$ let Φ be a symplectic action of the Lie group G with Lie algebra $\mathfrak{g}$. We say that a mapping $J : \mathbb{R}^{2n} \to \mathfrak{g}^*$ is a *momentum mapping for the action* Φ if for every $\xi \in \mathfrak{g}$ we have

$$X_{\hat{J}(\xi)} = \frac{d}{dt}\Phi(\exp t\xi,x)\Big|_{t=0}$$

where the right hand side is called the *infinitesimal generator* of the action corresponding to ξ. $\hat{J}(\xi) : \mathbb{R}^{2n} \to \mathbb{R}$ is defined by $\hat{J}(\xi)(x) = J(x).\xi$.

1.19. PROPOSITION. Let Φ be a symplectic action on $(\mathbb{R}^{2n},\omega)$ of the Lie group G having momentum mapping J. If G is a symmetry group for $(\mathbb{R}^{2n},\omega,H)$ then $\{\hat{J}(\xi),H\} = 0$.

If one considers a one-parameter symmetry group $\exp(t\ \mathrm{ad}(F))$, $t \in \mathbb{R}$ for $(\mathbb{R}^{2n},\omega,H)$ then one obtains a momentum mapping J such that $\hat{J}(\xi) : \mathbb{R}^{2n} \to \mathbb{R}\ ;\ x \mapsto F(x)$. Consequently F is an integral for $(\mathbb{R}^{2n},\omega,H)$.

Let G be a Lie group and $\mathfrak{g}$ its Lie algebra. If $g \in G$ then $I(g) : h \to ghg^{-1}$ is a isomorphism of G onto itself. Put $\mathrm{Ad}(g) = dI(g)_e$ then $\mathrm{Ad}(g)$ is an automorphism of $\mathfrak{g}$. We have $\mathrm{Ad}(\exp X) = \exp \mathrm{ad}(X)$ for $X \in \mathfrak{g}$. $\mathrm{Ad}^*(g)$ is the corresponding automorphism of $\mathfrak{g}^*$. Also $\mathrm{Ad}^*(g^{-1})$ is an automorphism of $\mathfrak{g}^*$, its action is called the co-adjoint action of G.

1.20. DEFINITION. We say that a momentum mapping J is Ad*-*equivariant* if $J(\Phi_g(x)) = \mathrm{Ad}^*(g^{-1})(J(x))$ for every $g \in G$.

It is clear that the momentum mapping for a one-parameter group $\exp(t\ \mathrm{ad}(F))$, $t \in \mathbb{R}$, $F \in C^\infty(\mathbb{R}^{2n},\mathbb{R})$ is trivially Ad*-equivariant.

Under certain conditions the presence of a symmetry group for a Hamiltonian system allows us to reduce our system to a system of lower dimension. With some abuse of language one might say that the reduced system is obtained by factoring out the symmetry group. We will state the classical reduction theorem as it can be found in Abraham and Marsden [1978] and Arnold [1978]. Our own construction of reduced systems in chapter 4 will be somewhat different.

1.21. THEOREM. Let G_X denote the isotropy subgroup of G under the coadjoint action Ad*, that is, $G_X = \{g \in G \mid \mathrm{Ad}^*(g^{-1})X=X\}$. Furthermore let G be a Lie group acting symplectically on $(\mathbb{R}^{2n},\omega)$ and let $J : \mathbb{R}^{2n} \to \mathfrak{g}^*$ be an Ad*-equivariant momentum mapping for this action. Assume $X \in \mathfrak{g}^*$ is a regular value for J and that G_X acts freely and properly on $J^{-1}(X)$. Then $M_X = J^{-1}(X)/G_X$ has a unique symplectic form ω_X defined by $\pi_X^*\omega_X = i_X^*\omega$ where $\pi_X : J^{-1}(X) \to M_X$ is the canonical projection and $i_X : J^{-1}(\mu) \to \mathbb{R}^{2n}$ is the inclusion. We call M_X the *reduced phase space*. Furthermore suppose that G is a symmetry group for $(\mathbb{R}^{2n},\omega,H)$. Then the flow φ of X_H leaves $J^{-1}(X)$ invariant and commutes with the action of G_X on $J^{-1}(X)$. Therefore there is a canonically induced flow φ_X on M_X which satisfies $\pi_X \circ \varphi = \varphi_X \circ \pi_X$. This flow is symplectic with a Hamiltonian function H_X which satisfies $H_X \circ \pi_X = H \circ i_X$. H_X is called the *reduced Hamiltonian function*. The system (M_X,ω_X,H_X) is called the *reduced Hamiltonian system*.

1.22. REMARK. The condition that X is a regular value is made in order to assure that $J^{-1}(X)$ is a manifold. However if X is not a regular value but part of $J^{-1}(X)$ is a manifold on which G_X acts freely and properly then one may still apply reduction to this part of $J^{-1}(X)$.

1.23. DEFINITION. A point $p \in \mathbb{R}^{2n}$ is called a *relative equilibrium* if its projection $\pi_X(p)$ onto the reduced phase space M_X is an equilibrium point for the reduced system (M_X,ω_X,H_X), where $X = J(p)$. (This is equivalent to saying that the G_X orbit is invariant under X_H).

1.24. DEFINITION. The map $H \times J : \mathbb{R}^{2n} \to \mathbb{R} \times \mathfrak{g}^*$ defined by the function H and the momentum map J is called the *energy-momentum mapping* for the system $(\mathbb{R}^{2n},\omega,H)$ with symmetry group G.

1.25. PROPOSITION. (i) $p \in J^{-1}(X)$ is a relative equilibrium if and only if p is a critical point of $H \times J$.

(ii) If $p \in J^{-1}(X)$ is a relative equilibrium then so is any element in the orbit of p under G given by $\mathcal{O}_G(p) = \{g.p | g \in G\}$.

As we have seen the existence of an integral F for a system $(\mathbb{R}^{2n},\omega,H)$ implies the existence of a one-parameter symmetry group. If this symmetry group fulfils the conditions of the reduction theorem 1.21. then the existence of an integral implies the possibility of constructing a reduced phase space. In general symmetries allow us to investigate the topology and geometry of Hamiltonian systems by considering reduced systems and energy-momentum maps. Because integrals and symmetries play such an important role we will conclude this section with some definitions and theorems concerning integrals and integrable systems.

1.26. <u>DEFINITION</u>. Two functions $F,G \in C^\infty(\mathbb{R}^{2n},\mathbb{R})$ are said to be in *involution* if $\{F,G\} = 0$. F and G are said to be *independent* if the set of points where dF and dG are linearly dependent has Lebesgue measure zero. The system $(\mathbb{R}^{2n},\omega,H)$ is called *integrable* if there are n integrals $F_i, i = 1,\ldots,n$ for this system such that F_i and F_j are in involution for all $i,j = 1,\ldots,n$ and such that F_i and F_j are independent for $i \neq j$, $i,j = 1,\ldots,n$.

For integrable systems we have the following basic theorem.

1.27. <u>THEOREM</u>. (see Arnold [1978]) Let $(\mathbb{R}^{2n},\omega,H)$ be an integrable Hamiltonian system with integrals F_i, $i = 1,\ldots,n$ as in definition 1.26. Suppose that the vector fields X_{F_i} are *complete*, that is, the flow of X_{F_i} is defined for all $t \in \mathbb{R}$. In addition suppose that on $M = \{x | F_i(x)=c_i\}$ the dF_i are independent at each point. Then M is an n-dimensional manifold whose connected components are T^n (if compact) or $\mathbb{R}^k \times T^{n-k}$. If M is a torus then in a neighbourhood of M there exist symplectic coordinates I_i,φ_i, $i = 1,\ldots,n$ such that the system

$(\mathbb{R}^{2n},\omega,H)$ takes the form

$$\frac{dI_i}{dt} = 0, \quad \frac{d\varphi_i}{dt} = w_i(I); \; i = 1,\dots,n$$

These coordinates are called *action-angle coordinates*. The action variables I_i are functions of the inegrals F_i.

In our applications (in two degrees of freedom, that is, $n = 2$) the integrals F and G will be polynomials. Then independence is equivalent to assuming that dF and dG are not everywhere linearly dependent, that is, the set where they are dependent is an algebraic subvariety of positive co-dimension. If dF and dG are linearly indpendent at every point of the surface M given by $F = c_1$ and $G = c_2$ then by theorem 1.27. M is a manifold consisting of tori and cylinders. Notice that M is precisely a regular fibre of the 'energy-momentum' map $F \times G$.

3. Linear Hamiltonian systems

As in section two consider $\mathbb{R}^{2n}$ with coordinates $(x,y) = (x_1,\dots,x_n,y_1,\dots,y_n)$ and standard symplectic form $\omega = \sum_{i=1}^{n} dx_i \wedge dy_i$. A Hamiltonian system $(\mathbb{R}^{2n},\omega,H)$ will be called linear if $X_H(x,y)$ is linear. In the following we will restrict ourselves to systems with Hamiltonian function H in the space of homogeneous quadratic polynomials P_2. Then $X_H(x,y) = A\binom{x}{y}$ with A a constant matrix. Moreover $A = DX_H(0)$. The set of linear symplectic maps from $\mathbb{R}^{2n}$ to $\mathbb{R}^{2n}$ form a group under composition called the *symplectic group* and denoted $Sp(n,\mathbb{R})$. Because $Sp(n,\mathbb{R}) \subset Gl(\mathbb{R}^{2n},\mathbb{R}^{2n})$ is a submanifold and composition is C^∞, $Sp(n,\mathbb{R})$ is a Lie group. The Lie algebra $sp(n,\mathbb{R})$ of $Sp(n,\mathbb{R})$ is called the algebra of *infinitesimally symplectic maps*. $sp(n,\mathbb{R})$ is a subalgebra of $gl(\mathbb{R}^{2n},\mathbb{R}^{2n})$ and has bracket $[A,B] = A\circ B - B\circ A$.

Notice that the Lie algebra $(sp(n,\mathbb{R}) : [\,,\,])$ is isomorphic to the Lie algebra $(P_2,\{\,,\,\})$ by the isomorphism $H \rightleftharpoons DX_H(0)$.

Choosing the standard basis in $\mathbb{R}^{2n}$ the matrix of ω is J given by (1.3). Note that $J^{-1} = J^t = -J$. Furthermore $J^2 = -I$. Relative to this basis A is a symplectic matrix if and only if $A^tJA = J$. B is an infinitesimal symplectic matrix if and only if $B^tJ + JB = 0$, that is, if and only if JB is symmetric.

Concerning the eigenvalues of infinitesimally symplectic maps we have the following.

1.28. PROPOSITION. Let $B \in sp(n,\mathbb{R})$. If λ is an eigenvalue of B of multiplicity k then so are $-\lambda, \bar{\lambda}$ and $-\bar{\lambda}$. If zero is an eigenvalue of B then it has even multiplicity.

The possible configurations for the eigenvalues of elements of $sp(2,\mathbb{R})$ are given in fig. (1.1) in the complex plane.

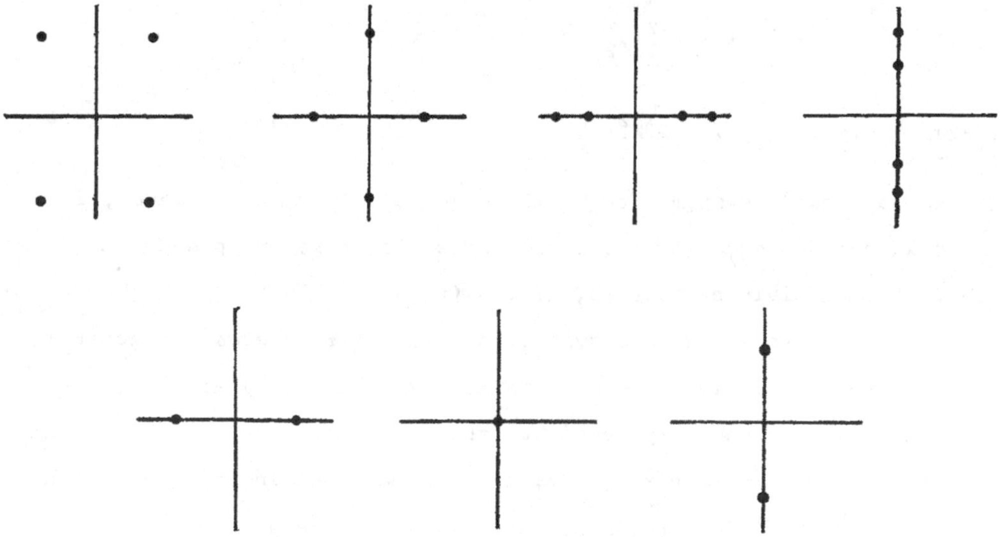

fig. (1.1). Configurations of the eigenvalues of $sp(2,\mathbb{R})$ in the complex plane

Of course one may normalize infinitesimal symplectic matrices with respect to conjugation by elements of $Sp(n,\mathbb{R})$. Restricting to the case of purely imaginary eigenvalues, in $sp(2,\mathbb{R})$ we get the normal forms given by (1.29) in the semisimple cases and by (1.30) in the nonsemisimple (and also non-nilpotent) cases (see Burgoyne and Cushman [1974]).

$$(1.29)\qquad \begin{pmatrix} 0 & 0 & \lambda_1 & 0 \\ 0 & 0 & 0 & \lambda_2 \\ -\lambda_1 & 0 & 0 & 0 \\ 0 & -\lambda_2 & 0 & 0 \end{pmatrix}$$

$$(1.30)\qquad \begin{pmatrix} 0 & -\alpha & 0 & 0 \\ \alpha & 0 & 0 & 0 \\ \varepsilon & 0 & 0 & -\alpha \\ 0 & \varepsilon & \alpha & 0 \end{pmatrix}$$

Here $\lambda_1,\lambda_2,\alpha$ may be positive or negative and $\varepsilon = \pm 1$. The corresponding quadratic Hamiltonian functions are

$$(1.31)\qquad \lambda_1(x_1^2+y_1^2) + \lambda_2(x_2^2+y_2^2)$$

$$(1.32)\qquad \alpha(x_1y_2-x_2y_1) - \tfrac{\varepsilon}{2}(x_1^2+x_2^2)$$

We say that a system with Hamiltonian (1.31) is in *resonance*, if λ_1 and λ_2 are dependent over the rationals. If there are p and q, in $\mathbb{Z}$ with no common divisor (that is, g.c.d. $(p,q) = 1$) such that $q\lambda_1 - p\lambda_2 = 0$ then we say the system is in $p : q$ *resonance*. In general we will take $p > 0$. Also systems with Hamiltonian (1.32) are said to be in resonance. Here one only considers the semisimple part of the Jordan decomposition of the matrix (1.30). It is clear that in this case we have a $1 : -1$ resonance. This resonance is called *non-semisimple* because of the presence of the nilpotent part in (1.30). Considering a system in resonance one might construct the unfolding (versal deformation) of the corresponding matrix (see Arnold [1971]). With exception of the semisimple 1 : 1 and 1 : -1 resonances, we get

$$(1.33)\qquad \begin{pmatrix} 0 & 0 & \lambda_1+\mu_1 & 0 \\ 0 & 0 & 0 & \lambda_2+\mu_2 \\ -\lambda_1-\mu_1 & 0 & 0 & 0 \\ 0 & -\lambda_2-\mu_2 & 0 & 0 \end{pmatrix}$$

and

$$(1.34)\qquad \begin{pmatrix} 0 & -\alpha-\nu_1 & \nu_2 & 0 \\ \alpha+\nu_1 & 0 & 0 & \nu_2 \\ \varepsilon & 0 & 0 & -\alpha-\nu_1 \\ 0 & \varepsilon & \alpha+\nu_1 & 0 \end{pmatrix}$$

for the unfolding of (1.29) and (1.30). Notice that unfolding is just detuning the resonance. For the unfolding of the semisimple 1 : 1 and 1 : -1 resonance one will need more then two parameters. For the semisimple 1 : -1 resonance one might take (1.34) as an unfolding considering ε as a parameter.

We shall treat the nonsemisimple case in a little more detail. We will work in the space P_2 of homogeneous quadratic polynomials on $\mathbb{R}^4$. Recall that ad(H) with H given by (1.32) is an endomorphism of P_2. Let Sp(2,$\mathbb{R}$) act on P_2 by composition. The tangent space to the orbit through H is given by im ad(H). Therefore, to determine the unfolding, we have to determine a complement to im ad(H). Such a complement C is given by the complement of im ad(X) in ker ad(S). Here $X = \frac{1}{2}(x_1^2+x_2^2)$ and $S = x_1y_2 - x_2y_1$. Also εX and αS correspond to the nilpotent resp. semisimple part of the Jordan decomposition of (1.30) (see van der Meer [1982]). Now ker ad(S) (ad(S) considered as an endomorphism of P_2) is just a Lie subalgebra of P_2 isomorphic to the centralizer of X_S in sp(2,$\mathbb{R}$) which in turn is isomorphic to u(1,1). Therefore ker ad(S) is spanned by S,X,$Y = \frac{1}{2}(y_1^2+y_2^2)$ and $Z = x_1y_1 + x_2y_2$. Moreover, X,Y and Z span a Lie subalgebra isomorphic to sl(2,$\mathbb{R}$), with X and Y corresponding to its nilpotent generators. The adjoint representation of ker ad(S) on $\mathcal{P}_2$ is isomorphic to $sl(2,\mathbb{R}) \times \mathbb{R}$ with ad(X) and ad(Y) corresponding to the nilpotent generators of $sl(2,\mathbb{R})$. Moreover we have

$$\mathcal{P}_2 = \ker ad(X) \oplus im\ ad(Y) = \ker ad(Y) \oplus im\ ad(X)$$

It follows that $C = \ker ad(S) \cap \ker ad(Y)$. Thus an unfolding of H is given by

$$H_\nu(x,y) = (\alpha+\nu_1)(x_1y_2-x_2y_1) - \frac{\varepsilon}{2}(x_1^2+x_2^2) + \frac{\nu_2}{2}(y_1^2+y_2^2)$$

which corresponds to (1.34). According to the sign of $\varepsilon\nu_2$ the eigenvalues of (1.34) are given in fig. (1.2).

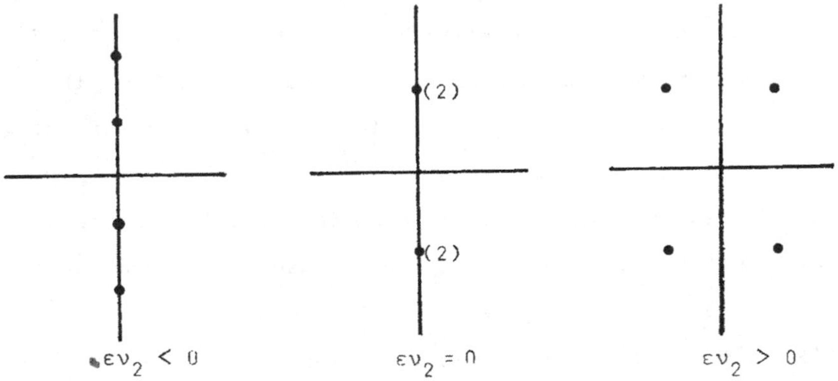

fig. (1.2) Eigenvalues of (1.34) with multiplicity.

Because changing the sign of $\varepsilon\nu_2$ from positive to negative gives the Hamiltonian version of passage of the eigenvalues "through" the imaginary axis, this resonance is referred to as the Hamiltonian Hopf bifurcation.

Consider a nonlinear system $(\mathbb{R}^{2n},\omega,H)$. Suppose that the Taylor series of H at zero is $H = H_2$ + (terms of order greater then two) with $H_2 \in \mathcal{P}_2$ then $(\mathbb{R}^{2n},\omega,H_2)$ is the linearized system corresponding to $(\mathbb{R}^{2n},\omega,H)$. In addition we say that $(\mathbb{R}^{2n},\omega,H)$ is in resonance if $(\mathbb{R}^{2n},\omega,H_2)$ is.

Chapter II

Normal forms for Hamiltonian functions

0. Introduction

In this chapter we will treat some aspects of the normalization of Hamiltonian systems. We will focus on the normalization of the Hamiltonian function. Clearly normalizing the Hamiltonian induces a normalization of the corresponding Hamiltonian vector field.

Let (M,ω) be a symplectic manifold. To each $F \in C^\infty(M,\mathbb{R})$ we associate a Hamiltonian system (M,ω,F) and a Hamiltonian vector field X_F. Furthermore we define the Poisson bracket $\{.,.\}$ as in definition 1.6., thus making $C^\infty(M,\mathbb{R})$ a Lie algebra. For $H \in C^\infty(M,\mathbb{R})$ define $ad(H)$ to be the map which assigns to each $F \in C^\infty(M,\mathbb{R})$ the bracket $\{H,F\}$.

Let G be the group of symplectic diffeomorphisms on M (see definition 1.5.). The action of G on M induces an action . on $C^\infty(M,\mathbb{R})$ given by $\varphi.F = F\circ\varphi$ for $\varphi \in G$, $F \in C^\infty(M,\mathbb{R})$. Because $\psi.\varphi.\ F = F\circ\varphi\circ\psi$, the group acting on $C^\infty(M,\mathbb{R})$ is antihomomorphic to G. However by proposition 1.15.(iii) this is a natural action. With abuse of language we will speak of the action of G on $C^\infty(M,\mathbb{R})$. In fact the action of $\varphi \in G$ on $C^\infty(M,\mathbb{R})$ is pull-back, that is, for $F \in C^\infty(M,\mathbb{R})$ $\varphi.F = \varphi^*F$. Therefore the action of each $\varphi \in G$ on $C^\infty(M,\mathbb{R})$ is a Lie algebra isomorphism.

Let $O_G(F)$ be the orbit of F under the action of G. Let $N(K)$ be a space complementary to the tangent space to $O_G(K)$ at K. The following generalizes the classical definition of Hamiltonian normal form.

2.1. DEFINITION. Let $H,K,F \in C^\infty(M,\mathbb{R})$. H is a K-*normal form* for F if $H \in O_G(F) \cap N(K)$.

Notice that this definition can be generalized to elements of an arbitrary vector space V, given a group action on V.

From definition 2.1. it is clear that one has to choose K and $N(K)$

before one starts normalization. Furthermore it is clear that, with an arbitrary choice of K, a K-normal form for a given function F will in general not exist. Classically one considers normalization of the Taylor series of a Hamiltonian function at an isolated critical point. Such a Taylor series starts with quadratic terms. For K one takes the homogeneous quadratic terms in the Taylor-series. If K is semisimple $N(K)$ can be chosen as the centralizer of K in $(C^\infty(\mathbb{R}^{2n},\mathbb{R}),\{,\})$. Even in this case a K-normal form is in general not convergent (see Siegel [1952], Brjuno [1971]), that is, a K-normal form in the sense of the definition does not exist. However, one may develop a formal theory, which also shows how to normalize Taylor series up to arbitrary order.

In the following we will restrict ourselves to the case of formal power series Hamiltonians. This is in the following way related to normalization up to arbitrary order of Taylor series of Hamiltonians at $p \in M$. Let $E(M,p)$ be the algebra of germs of C^∞-functions on M at p. Define $E_k^+ = \{F \in E(M,p) \mid F(z) = O(|z|^k), z \to p\}$ and $E_\infty = \bigcap_{k\geq 0} E_k^+$. Notice that we have $E_k^+ . E_m^+ \subset E_{k+m}^+$ considering $E(M,p)$ as an algebra, while $\{E_k^+, E_m^+\} \in E_{k+m-2}^+$ using the Lie algebra structure. Now let $P = E(M,p)/E_\infty$ be the space of formal power series. By a theorem of Borel every formal power series, in local coordinates, is the Taylor series of an C^∞-function. By the theorem of Darboux (theorem 1.6.) in a neighborhood of $p \in (M,\omega)$ we have coordinates $(x,y) = (x_1,\ldots,x_n,y_1,\ldots,y_n)$ such that $p = (0,0)$ and $\omega = \sum_{i=1}^{n} dx_i \wedge dy_i$. We will use these local coordinates. Then $P_k^+ = E_k^+/E_\infty$ is the space of formal power series starting with terms of order k and $P_k^- = P/P_{k+1}^+$ is the space of polynomials of order $\leq k$. Furthermore $P_k = P_k^+/P_{k+1}^+$ is the space of homogeneous polynomials of order k. Notice that $P_k^- \cong E/E_{k+1}^+$ and $P_k \cong E_k^+/E_{k+1}^+$. Thus P_k^- is the space of k-jets of germs in $E(M,p)$, that is, in local coordinates P_k^- is the space of k-jets of Taylor series at zero of functions in

$C^\infty(\mathbb{R}^{2n},\mathbb{R})$. Thus normalizing k-jets of formal power series corresponds to normalizing k-jets of Taylor series (or k-jets of germs).

We now restrict ourselves to looking at formal power series in P_2^+; *henceforth let* $P = P_2^+ = \sum_{k=2}^{\infty} P_k$. Under Poisson bracket P is a Lie algebra isomorphic to the Lie algebra of Hamiltonian formal power series vector fields with the origin as equilibrium. *Furthermore let G denote the group of invertible origin preserving symplectic formal power series transformations on* $\mathbb{R}^{2n}$. As before G induces an action on P by composition.

We now summarize the contents of this chapter. In the first section we treat S-normal forms for $H \in P$, where S is the semisimple part in the Jordan-Chevalley decomposition of the quadratic part H_2 of H. Notice that we may write $H = \sum_{k=2}^{\infty} H_k$ with $H_k \in P_k$, $k \geq 2$. We show how to determine $N(S)$. Furthermore we show that $O_G(H) \cap N(S)$ is the orbit of an S-normal form for H under some subgroup of G. Thus in general S-normal forms are not unique.

In the second section we study the orbit of $H \in P$ under G. We show that further normalization of S-normal forms is possible, although it is hard to perform. In particular we find H_2-normal forms for $H \in P$, including the case when H_2 has a nontrivial nilpotent part in its Jordan-Chevalley decomposition.

In the third section we give several examples which illustrate the theory of sections 1 and 2.

In section 4 we study what normal form theory implies about the existence of integrals and the existence of Hamiltonian normal forms for energy-momentum mappings.

Finally in section 5 we give a brief historical review of normal form theory. An extensive bibliography follows the main lines in the development of normal form theory. Because of the large number of publications on this subject, this bibliography is not complete.

1. Normalization with respect to the semisimple part of H_2

Let $H \in P$. Then according to definition 2.1. the process of finding a normal form for H leads to making two choices. First we have to choose K and second we have to choose $N(K)$

As is well known the tangent space T_K to the G-orbit of K at K is given by $T_K = K + \{K,P\} = K + \text{imad}(K)$. Therefore $N(K)$ may be chosen as a complement of imad(K) in P at K. Choose $N(K) = K + \tilde{N}(K)$ where $\tilde{N}(K)$ is defined by $\tilde{N}(K) \oplus \text{imad}(K) = P$. With this choice we have the following lemmas.

2.2. LEMMA. If H is a K-normal form for F then $H\circ\varphi$ is a $K\circ\varphi$-normal form for F.

Proof. For $\varphi \in G$ we have $\text{imad}(K\circ\varphi) = \varphi.(\text{imad}(K))$ because $\text{imad}(K\circ\varphi) =$ $= \text{ad}(K\circ\varphi)P = \{K\circ\varphi,P\} = \varphi.\{K,\varphi^{-1}.P\} = \varphi.\{K,P\}$ because φ is a Lie algebra isomorphism. If we choose $N(K\circ\varphi)$ to be $\varphi.N(K)$ then $\tilde{N}(K\circ\varphi) \oplus \text{imad}(K\circ\varphi) = P$. Because $H \in N(K) \cap O_G(F)$ it follows that $H\circ\varphi \in N(K\circ\varphi) \cap O_G(F)$. ⊠

The following lemma shows that, in general, normal forms are not unique.

2.3. LEMMA. Let φ be in the subgroup of G which leaves K invariant, that is, $\varphi.K = K$, then $H\circ\varphi$ is a K normal form for F if and only if H is a K-normal form for F.

Proof. Follows directly from lemma 2.2.

We now choose K. Recall that P_2 with Poisson bracket is isomorphic to the semisimple Lie algebra $sp(n,\mathbb{R})$ (see ch.1, sect.3). Moreover the Jordan-Chevalley decomposition (hereafter the S - N decomposition) of elements in $sp(n,\mathbb{R})$ carries over to elements of P_2, that is, for $H_2 \in P_2$ there is a unique decomposition $H_2 = S + N$ with S semisimple

N nilpotent and $\{N,S\} = 0$. *For the rest of this section we will only consider* S*-normal forms for* $H \in P$ where S is the semisimple part of H_2. Next we give a characterization of $N(S)$ which in general allows us to show what the normal form will look like (see the examples in section 3). Let $ad_m(F)$ denote the restriction of ad(F) to P_m.

2.4. LEMMA. A convenient choice for $N(S)$ is kerad(S).

Proof. Now $ad_2(P_2)$ is the adjoint representation of P_2 and thus is isomorphic to P_2. Besides that, every $ad_m(P_2)$ is a representation of the semisimple Lie algebra P_2. By Humphreys [1972] corollary 6.4. the S - N decomposition of P_2 carries over to the representations $ad_m(P_2)$, that is, $ad_m(H_2) = ad_m(S) + ad_m(N)$ is the S - N decomposition of $ad_m(H_2)$. Thus $ad_m(S)$ is semisimple for $m \geq 2$. Therefore $P_m = \text{kerad}_m(S) \oplus \text{imad}_m(S)$. Now $\text{imad}(S) = \sum_{m=2}^{\infty} \text{imad}_m(S)$ and $\text{kerad}(S) = \sum_{m=2}^{\infty} \text{kerad}_m(S)$ because ad(S) acts linearly on P and preserves the degree. Hence imad(S) as well as kerad(S) has a homogeneous basis. Since $\text{kerad}(S) \oplus \text{imad}(S) = P$, we may choose $\tilde{N}(S) = \text{kerad}(S)$. But then $N(S) = S + \tilde{N}(S) = \text{kerad}(S)$. ⊠

The following algorithm shows how $H \in P$ can be transformed by successive symplectic transformations into S-normal form up to arbitrary order. Here H being in S-normal form up to order k means that $\sum_{m=2}^{k} H_m$ is in S-normal form. Consequently we may formally transform H into S-normal form (up to infinite order) by an infinite succession of symplectic maps.

2.5. THEOREM. Let $F \in P$ and let $S \neq 0$ be the semisimple part of F_2. Then for each $k \in \mathbb{N}$, $k \geq 2$ there exists a $\varphi \in G$ such that $H = F \circ \varphi$ is in S-normal form up to order k.

proof. Clearly F_2 is in S-normal form because $F_2 = S + N$ with $N \in \text{kerad}(S)$. Suppose that F is in normal form up to order $k - 1$. Then for $P \in \mathcal{P}_k$ $\operatorname{expad}(P)\,F = \sum_{m=2}^{k-1} F_m + F_k + \{P,F_2\}$ + (terms of order greater then k). Because $\mathcal{P}_k = \operatorname{kerad}_k(S) \oplus \operatorname{imad}_k(S)$ we may write $F_k = \hat{F}_k + \tilde{F}_k$ where $\hat{F}_k \in \operatorname{kerad}_k(S)$ and $\tilde{F}_k \in \operatorname{imad}_k(S)$. Similarly we have a splitting for $P = \hat{P} + \tilde{P}$. This means that we can write the k-th order term of $\operatorname{expad}(P)F$ as $\hat{F}_k + \tilde{F}_k + \{\hat{P},N\} + \{\tilde{P},F_2\}$. Now $\{\hat{P},N\} \in \operatorname{im}(\operatorname{ad}_k(N)|\operatorname{kerad}_k(S)) = \operatorname{imad}_k(N) \cap \operatorname{kerad}_k(S)$. Thus $\{\hat{P},N\} \in \operatorname{kerad}_k(S)$. Furthermore $\{\tilde{P},F_2\} \in \operatorname{im}(\operatorname{ad}_k(F_2)|\operatorname{imad}_k(S)) = \operatorname{imad}_k(S)$. (see the proof of lemma 2.2, van der Meer [1982]). Therefore we may choose P so that $\tilde{F}_k + \{\tilde{P},F_2\} = 0$. The remaining k-th order terms then are $\hat{F}_k + \{\hat{P},N\}$ which are in $\operatorname{kerad}_k(S)$. Thus $F \circ \operatorname{expad}(P)$ is in S-normal form up to order k. By induction the theorem follows (cf. Chen [1963] prop. 8.1; Takens [1974], Th. 2.1; Broer [1979] Th. 2.3.6). ⊠

From the proof of the above theorem it is clear that one will need a formal transformation of the form $\operatorname{expad}(P_3)\circ\operatorname{expad}(P_4)\circ\ldots\circ\operatorname{expad}(P_k)\circ\ldots$ to put F in S-normal form. Notice that at each step of the above algorithm expad(P) is determined up to terms of P in $\operatorname{kerad}_k(S)$. Starting with $k = 3$ this freedom of choice might lead to different normal forms up to order greater then three. In the following we will show that these normal forms can be transformed into each other. We need the next lemma.

2.6. <u>LEMMA</u>. Let S be the semisimple part of H_2. Then for $P_k \in \mathcal{P}_k, k \geq 2$ $\{H_2,P_k\} \in \operatorname{kerad}(S)$ if and only if $P_k \in \operatorname{kerad}(S)$.

<u>proof</u>. Let $P_k = \hat{P}_k + \tilde{P}_k$, with $\hat{P}_k \in \operatorname{kerad}_k(S)$ and $\tilde{P}_k \in \operatorname{imad}_k(S)$, be the splitting of P_k corresponding to $\mathcal{P}_k = \operatorname{kerad}_k(S) \oplus \operatorname{imad}_k(S)$. Then $\{H_2,P_k\} = \{H_2,\tilde{P}_k\} + \{N,\hat{P}_k\}$. If $P_k \in \operatorname{kerad}(S)$ then $\tilde{P}_k = 0$ and obviously $\{H_2,P_k\} \in \operatorname{kerad}(S)$; which proves the "if" part. Now suppose $\{H_2,P_k\} \in \operatorname{kerad}_k(S)$. Because $\{N,\hat{P}_k\} \in \operatorname{kerad}_k(S)$ we have

$\{H_2,\tilde{P}_k\} \in \ker ad_k(S)$. Furthermore $\mathrm{im}(ad_k(H_2)|\mathrm{im}\,ad_k(S)) = \mathrm{im}\,ad_k(S)$ which means that $\{H_2,\tilde{P}_k\} \in \mathrm{im}\,ad_k(S) \cap \ker ad_k(S)$, that is, $\{H_2,\tilde{P}_k\} = 0$. But $\ker ad_k(H_2) = \ker ad_k(S) \cap \ker ad_k(N)$. Hence $ad_k(S)\tilde{P}_k = \{S,\tilde{P}_k\} = 0$. Thus $\tilde{P}_k \in \ker ad_k(S) \cap \mathrm{im}\,ad_k(S)$, that is, $\tilde{P}_k = 0$. Thus $P_k = \hat{P}_k \in \ker ad_k(S) \subset \ker ad(S)$. ⊠

2.7. THEOREM. Let $P \in \mathcal{P}_3^+$ and let $H \in \ker ad(S)$. Then $\exp ad(P).H \in \ker ad(S)$ if and only if $P \in \ker ad(S)$.

proof. If $P \in \ker ad(S)$ then $\exp ad(P).S = S$. Consequently if $\{S,H\} = 0$ then $\{S,\exp ad(P).H\} = 0$, that is, $\exp ad(P).H \in \ker ad(S)$. Next suppose that $\exp ad(H).P \in \ker ad(S)$. Let $\pi_m : \mathcal{P} \to \mathcal{P}_m$ be the projection of elements of $\mathcal{P}$ onto their m-th order term. Then $\exp ad(P).H \in \ker ad(S)$ is equivalent to $\pi_m(\exp ad(P).H) \in \ker ad_m(S)$ for all $m \geq 3$. We will prove by induction that $P_m \in \ker ad_m(S)$ for all $m \geq 3$. First consider $\pi_3(\exp ad(P)H\} =$ $= H_3 + \{P_3,H_2\}$. Because $\pi_3(\exp ad(P).H) \in \ker ad_3(S)$ and $H_3 \in \ker ad_3(S)$ we have $\{P_3,H_2\} \in \ker ad_3(S)$ and thus by lemma 2.6. $P_3 \in \ker ad_3(S)$. Now suppose that $P_k \in \ker ad_k(S)$ for $3 \leq k \leq m-1$ then $\pi_m(\exp ad(P).H) =$ $= H_m + \{P_m,H_2\}$ + (brackets of elements in $\ker ad(S)$), hence $\{P_m,H_2\} \in$ $\in \ker ad_m(S)$. Thus $P_m \in \ker ad_m(S)$. By induction $P \in \ker ad(S)$. ⊠

The normal form algorithm (theorem 2.5.) only uses transformations of the form $\exp ad(P)$, $P \in \mathcal{P}_3^+$. If we want to normalize further using only maps of the form $\exp ad(P)$, $P \in \mathcal{P}_3^+$, then by theorem 2.7. we can perform further normalization inside $\ker ad(S)$. This approach does not use the full action of G. However, we can formulate an analogue of theorem 2.7. considering the full action of G. Before doing so we need a more detailed description of the Lie algebra structure of $\mathcal{P}$ and the Lie group structure of G.

Consider $\mathcal{P}_{m+1}^+$. Because $\{\mathcal{P}_k,\mathcal{P}_l\} \subset \mathcal{P}_{k+l-2}$ we see that $\mathcal{P}_{m+1}^+$ is an ideal in the Lie algebra $\mathcal{P}$. Thus $\mathcal{P}/\mathcal{P}_{m+1}^+ \cong \mathcal{P}_m^-$ is a Lie algebra. The Lie bracket on $\mathcal{P}_m^-$ is just the restriction of the Poisson bracket on $\mathcal{P}$.

Let $j_m : \mathcal{P} \to \mathcal{P}_m^-$ be the projection of elements of $\mathcal{P}$ onto their m-jets. Then on $\mathcal{P}_m^-$ we have the Lie bracket $j_m\{.,.\}$. (cf. Sternberg [1961], sect. 1). For $P \in \mathcal{P}_m^-$ let $ad_{j_m}(P)$ be the restriction of $j_m \circ ad(P)$ to $\mathcal{P}_m^-$. Then $ad_{j_m}(P) : \mathcal{P}_m^- \to \mathcal{P}_m^-$ and ad_{j_m} is the restriction to P_m of the adjoint representation of P. For $m \to \infty$ we have $ad_{j_m} \to ad$.

2.8. DEFINITION. $P \in \mathcal{P}_m^-$ is semisimple (resp. nilpotent) if $ad_{j_m}(P) : \mathcal{P}_m^- \to \mathcal{P}_m^-$ is semisimple (resp. nilpotent).

2.9. DEFINITION. $P \in \mathcal{P}$ is semisimple (resp. nilpotent) if $j_m P$ is semisimple (resp. nilpotent) in $\mathcal{P}_m^-$ for all m.

2.10. REMARK. Notice that each $P \in \mathcal{P}_m^- \cap \mathcal{P}_3^+$ is nilpotent in $\mathcal{P}_m^-$. Thus each $P \in \mathcal{P}_3^+$ is nilpotent in $\mathcal{P}$. It easily follows that $P \in \mathcal{P}$ is nilpotent if and only if P_2 is nilpotent or zero. Consequently if $P \in \mathcal{P}$ is semisimple then P_2 is nonzero and semisimple.

Consider the action of G on $\mathcal{P}_m^-$ given by $\varphi.P = j_m(P \circ \varphi)$, $P \in \mathcal{P}_m^-$, $\varphi \in G$. (Which is the action of G on $\mathcal{P}$ restricted to $\mathcal{P}_m^-$ and then truncated at order greater then m). It is clear that for this action only the m-jet of $\varphi \in G$ is of any importance. Therefore let G_m^- be the space of formal power series transformations truncated after order m. If composition in G is followed by truncation then G_m^- becomes a transformation group. (cf. Sternberg [1961], sect. 1)

If we consider $\mathcal{P}_m^-$ as a Lie algebra then each $P \in \mathcal{P}_m^-$ has a S - N decomposition corresponding to the S - N decomposition of $ad_{j_m}(P)$. By a transformation in G_m^- each $F \in \mathcal{P}_m^-$ can be transformed to an S_2-normal form H, where S_2 is the nonzero semisimple part of $F_2 = H_2$, or is nilpotent by remark 2.10. When F is nilpotent its S - N decomposition is trivial. When F_2 (and thus F) is not nilpotent, we consider a S_2-normal form H of F. Clearly $H = S_2 + N$, with $N = N_2 + \sum_{k=3}^{m} H_2$, and $\{S_2, N\} = 0$.

Thus $H = S_2 + N$ is the S - N decomposition of $H \in P_m^-$. Because the transformation to the normal form H is a Lie algebra isomorphism we have an S - N decomposition for F. Letting m go to infinity we obtain a S - N decomposition for elements of P. More precisely, each $P \in P$ is formally conjugate in G to an $H \in P$ such that the S - N decomposition of H is $S_2 + N$ with S_2 the semisimple part of $H_2 = P_2$ and $\pi_2 N = N_2$ the nilpotent part of F_2. If the semisimple part S_2 of P_2 is zero then P is nilpotent. (cf. Chen [1963], th. 8.1).

Next consider G_m^- as a finite dimensional transformation group. For each $\varphi \in G_m^-$ there exists a unique decomposition $\varphi = \varphi_s \circ \varphi_u$ with φ_s semisimple and φ_u unipotent. Furthermore for each unipotent φ_u there exists a nilpotent $N \in P_m^-$ such that $\varphi_u = \exp \mathrm{ad}_{j_m}(N)$. Letting m go to infinity we obtain a semisimple-unipotent decomposition for $\varphi \in G$. Moreover each unipotent $\varphi \in G$ can be written as $\varphi = \exp \mathrm{ad}(N)$ with $N \in P$ nilpotent.

2.11. <u>LEMMA</u>. Any map $\varphi \in G$ can be written as $\mathrm{expad}(P_1) \circ A$ or $A \circ \mathrm{expad}(P_2)$ with $A \in Sp(n,\mathbb{R})$ invertible and $P_1, P_2 \in P_3^+$.

<u>proof</u>. Because φ is invertible and origin preserving its linear part A is invertible and thus $\varphi = \psi_1 \circ A$ or $\varphi = A \circ \psi_2$ with $\psi_1 = \varphi \circ A^{-1}$ and $\psi_2 = A^{-1} \circ \varphi$. If we consider φ as a map from $\mathbb{R}^{2n}$ to $\mathbb{R}^{2n}$ then A is a $2n \times 2n$ symplectic matrix since φ is a symplectic diffeomorphism. In addition ψ_1 and ψ_2 are unipotent with linear part the identity. Therefore ψ_1 and ψ_2 can be written as $\mathrm{expad}(P_1)$ and $\mathrm{expad}(P_2)$ with P_1 and P_2 in P_3^+. Notice that because $A \circ \mathrm{expad}(P) \circ A^{-1} = \mathrm{expad}(P \circ A)$ we have $P_1 \circ A = P_2$.

Lemma 2.11. allows us to determine the full group of transformations mapping S-normal forms to S-normal forms. In fact we can consider linear mappings and mappings of the form $\mathrm{expad}(P)$, $P \in P_3^+$ seperately when dealing with the action of G.

2.12. THEOREM. Suppose H is an S_2-normal form for $F, H_2 = F_2$, where S_2 is the semisimple part of $H_2 = F_2$. Then $H \circ \varphi$ is an S_2-normal form for F if and only if $\varphi = A \circ \mathrm{expad}(P_1) = \mathrm{expad}(P_2) \circ A$ where A is a linear symplectic map such that $H_2 \circ A = H_2$ and $P_1, P_2 \in P_3^+ \cap \mathrm{kerad}(S_2)$.

proof. If we want to normalize $F \in P$ we may suppose F_2 to be in normal form. F_2 remains unchanged during the normalization procedure. Therefore any map mapping a S_2-normal form H to an S_2-normal form must leave $F_2 = H_2$ fixed. For φ only A acts on H_2 in P_2. Thus we must have $H_2 \circ A = H_2$ for $H \circ \varphi$ to be an S_2-normal form again.

When $H_2 \circ A = H_2$ we also have $S_2 \circ A = S_2$. Thus A is an isomorphism of $\mathrm{kerad}(S_2)$. Therefore $\mathrm{expad}(P_1)$ as well as $\mathrm{expad}(P_2)$ map $H \in \mathrm{kerad}(S_2)$ to an element of $\mathrm{kerad}(S_2)$. By theorem 2.7. we have P_1 and $P_2 \in P_3^+ \cap \mathrm{kerad}(S_2)$. This proves the "only if part".

The "if" part follows immediately if one realizes that if $H_2 \circ A = H_2$ then A is an isomorphism of $\mathrm{kerad}(S_2)$. Now use theorem 2.7. ⊠

2.13. REMARK. Suppose that H is an S_2-normal form for $F \in P$. Let $G^{(2)}_{H,S_2} = \{\varphi \in G \mid \varphi = A \circ \mathrm{expad}(P),\ H_2 \circ A = H_2,\ P \in P_3^+ \cap \mathrm{kerad}(S_2)\}$ with S_2 the semisimple part of H_2. Then by theorem 2.12. all S_2-normal forms for F are in the orbit of H under $G^{(2)}_{H,S_2}$. That is, for studying further normalization of H we may restrict to studying the orbit structure of $P_3^+ \in \mathrm{kerad}(S_2)$ under the action of $G^{(2)}_{H,S_2}$. (Note that $G^{(2)}_{H,S_2}$ is just the group of formal power series transformations equivariant with respect to the flow of X_{S_2} and preserving the two-jet of H).

2.14. REMARK. When for $F \in P$ $S_2 = 0$, that is, when F is nilpotent then F is trivially in S_2-normal form because $\mathrm{kerad}(S_2) \cap P_3^+ = P_3^+$. Thus in this case $G^{(2)}_{H,S_2} = G^{(2)}_F$ the group of transformations φ in G that preserve the two-jet of F, that is, $j_2F = j_2(F \circ \varphi)$.

2. Further normalization

Before going into the further normalization of S-normal forms we will make some general remarks about the structure of the orbit of H under the action of G. First we consider what happens to H_2. It is clear that for this we have only to look at the linear parts of G, that is, G_1^- (or j_1G) which is just the linear symplectic group. Thus the change of H_2 in P_2 is just given by its orbit under the linear symplectic group. Fixing an element of the orbit of H_2 in P_2, we then consider the group $G_H^{(2)} = \{\varphi \in G \mid j_2(\varphi.H) = j_2H\}$. We are now interested in what happens to H_3 under the action of $G_H^{(2)}$ restricted to P_3, that is, in $\pi_3(G_H^{(2)}.H)$. Notice that $\pi_3(G^{(2)}.H) = \pi_3(G_H^{(2)}.j_3H) = \pi_3(j_2G_H^{(2)}.j_3H)$ where $j_2G_H^{(2)} \subset G_2^-$. By a standard argument of group actions commuting with projections, we see that G.H is a fibre bundle with base $\pi_2(G.H)$ and fibre $G_H^{(2)}.H$. Next consider $G_H^{(3)} = \{\varphi \in G \mid j_3(\varphi.H) = j_3H\}$. In the same way we obtain $G_H^{(2)}.H$ as a fibre bundle with base $\pi_3(G_H^{(2)}.H)$ and fibre $G_H^{(3)}.H$. Thus we obtain an iterated fibre bundle

$$\pi_2(G.H) \leftarrow \pi_3(G_H^{(2)}.H) \leftarrow (G_H^{(3)}.H).$$

Of course we may extend this construction to arbitrary order. Let $G_H^{(m)} = \{\varphi \in G \mid j_m(\varphi.H) = j_mH\}$ and let $\pi_m(G_H^{(m)}.H)$ be *the orbit of* H_m *in* P_m. Then the orbit of H under G is given as the iterated fibre bundle

$$\pi_2(G.H) \leftarrow \pi_3(G_H^{(2)}.H) \leftarrow \dots \leftarrow \pi_{m-1}(G_H^{(m-1)}.H) \leftarrow (G_H^{(m)}.H)$$

At each step the fibre is just the orbit of H_k in P_k. Each $G_H^{(m)}$ is a subgroup of G with $G \supset G_H^{(2)} \supset G_H^{(3)} \supset \dots \supset G_H^{(m-1)} \supset G_H^{(m)}$. By lemma 2.11. $\varphi \in G_H^{(m)}$ can be written as $A \circ \mathrm{expad}(P)$ with A the linear part of φ.

2.15. <u>LEMMA</u>. There exists a $P \in P_3^+$ such that $A \circ \mathrm{expad}(P) \in G_H^{(m)}$ if and only if $j_m((A \circ \mathrm{expad}(P_3^+)).H) = j_m(\mathrm{expad}(P_3^+).H)$.

proof. In the right hand side of the last expression take the identity in $\mathrm{expad}(P_3^+)$. Then there exists a $P \in P_3^+$ such that $j_m((A\circ\mathrm{expad}(P)).H) = j_m H$, which proves sufficiency. Next suppose that $j_m((A\circ\mathrm{expad}(P)).H) = j_m H$. Rewriting the left hand side, using the action defined above, we get $j_m(\mathrm{expad}(P).(H\circ A)) = j_m H$. Now apply the action of $\mathrm{expad}(P_3^+)$ to both sides. Then we get $j_m(\mathrm{expad}(P_3^+).(H\circ A)) = j_m(\mathrm{expad}(P_3^+).H)$ which can be written as $j_m((A\circ\mathrm{expad}(P_3^+)).H) = j_m(\mathrm{expad}(P_3^+).H)$ ⊠

An immediate consequence of lemma 2.15. is

2.16. COROLLARY. The linear parts of $G_H^{(m)}$ (that is, $j_1 G_H^{(m)}$) form a subgroup $A_H^{(m)}$ of $Sp(n,\mathbb{R})$. Moreover if $k > m$ then $A_H^{(m)} \supset A_H^{(k)}$.

This means that we can construct an infinite descending series of subvarieties of $Sp(n,\mathbb{R})$. Since $Sp(n,\mathbb{R})$ is finite dimensional, and considered as an algebraic variety has only finitely many components, after finitely many steps our descending series of subvarieties $A_H^{(m)}$ must end up as some $A_H^{(m)}$. More precisely,

2.17. PROPOSITION. There exists a $m_0 \in \mathbb{N}$ such that for each $m \geqslant m_0$ $A_H^{(m)} = A_H^{(m_0)}$.

Using lemma 2.15. it follows that for $m \geqslant m_0$ the action of $G_H^{(m)}$ is equal to the action of $\{\mathrm{expad}(P) \mid P \in P_3^+, j_m(\mathrm{expad}(P).H) = j_m H\}$. We will consider this action now in a little more detail.

2.18. THEOREM. If $P \in P_3^+$ then $\mathrm{expad}(P).H = H$ if and only if $\{P,H\} = 0$.

proof. If $\mathrm{expad}(P).H = H$ then $\sum_{k=1}^{\infty} \frac{1}{k!}\mathrm{ad}^k(P)H = 0$. Thus $\left(\sum_{k=0}^{\infty} \frac{1}{(k+1)!}\mathrm{ad}^k(P)\right)(\mathrm{ad}(P)H) = 0$. Now $\sum_{k=0}^{\infty} \frac{1}{(k+1)!}\mathrm{ad}^k(P)$ is an invertible operator. Thus $\mathrm{ad}(P)H = \{P,H\} = 0$. When $\{P,H\} = 0$ trivially $\mathrm{expad}(P).H = H$. ⊠

2.19. <u>COROLLARY</u>. If $P \in \mathcal{P}_3^+$ then $j_m(\text{expad}(P).H) = j_m H$ if and only if $j_m\{P,H\} = 0$.

Notice that $\pi_m\{P,H\} = 0$ is equivalent to $\sum_{k+l=m+2} \{P_k,H_l\} = 0$. As a consequence of corollary 2.19. we have the following proposition which describes the orbit of H_{m+1}, $m \geq m_0$ in $\mathcal{P}_{m+1}$.

2.20. <u>PROPOSITION</u>. If $P \in \mathcal{P}_3^+$ and $j_m(\text{expad}(P).H) = j_m H$ then $j_{m+1}(\text{expad}(P).H) = j_{m+1}H + \pi_{m+1}\{P,H\}$.

<u>proof</u>. By corollary 2.19. $j_m(\text{ad}(P)H) = 0$ thus $j_{m+1}(\text{ad}^2(P)H) = 0$. Now $\sum_{k=0}^{\infty} \frac{1}{(k+2)!} \text{ad}^k(P)$ is an invertible operator on $\mathcal{P}_{m+1}^-$. Thus $j_{m+1}[(\sum_{k=0}^{\infty} \frac{1}{(k+2)!} \text{ad}^k(P))\text{ad}^2(P)H] = 0$. Therefore $j_{m+1}[\sum_{k=2}^{\infty} \frac{1}{k!} \text{ad}^k(P)H] = 0$ which is equivalent to $j_{m+1}((\text{expad}(P) - I - \text{ad}(P))H) = 0$. Thus $j_{m+1}(\text{expad}(P)H) = j_{m+1}H + j_{m+1}\{P,H\} = j_{m+1}H + \pi_{m+1}\{P,H\}$. ⊠

Notice that proposition 2.20. shows that the orbit of H_{m+1} in P_{m+1} for $m \geq m_0$ is a linear variety. Therefore, for $m \geq m_0$, we see that the fibre bundle $\pi_m(G_H^{(m)}.H) \leftarrow \pi_{m+1}(G_H^{(m+1)}.H)$ is an affine bundle. For $m < m_0$ the fibre will in general be some semialgebraic variety. Thus the fibre bundle will be far more complicated.

The above description of the orbit of $H \in \mathcal{P}$ under the action of G shows that normalization up to order k is nothing else than successively choosing an element in the orbit of H_m in $\mathcal{P}_m$ for each $m = 2,\ldots,k$. Depending on the criteria one uses, this choice is in general not uniquely determined; furthermore the choice at each level depends on the choices made on lower levels. If one wants to transform away as many terms as possible then one has to determine the invariants of the action of G (restricted) on $\mathcal{P}_m$, or in other words determine the orbits of the H_m in $\mathcal{P}_m$. The examples below show that this is more easily said than done.

The situation is simplified if one only considers the subgroup of G consisting of all expad(P), $P \in P_3^+$ (or $P \in P_3^+ \cap \mathrm{kerad}(S_2)$ if one starts with an S_2-normal form). By proposition 2.20. the orbit of H under this subgroup is an iterated fibre bundle which on each level is an affine bundle. Therefore the orbits on each level are more easy to determine.

Let $P_H^{(m)} = \{P \in P_3^+ | j_m\{P,H\} = 0\}$. Then by proposition 2.20. the orbit of H_{m+1} in P_{m+1} under $\mathrm{expad}(P_H^{(m)})$ is $H_{m+1} + \pi_{m+1}\,\mathrm{ad}(j_m H)(P_H^{(m)})$. Thus one may choose $P \in P_H^{(m)}$ so that H_{m+1} consist of terms lying in some complement of $\pi_{m+1}\,\mathrm{ad}(j_m H)(P_H^{(m)})$, that is, H_{m+1} is put into $j_m H$-normal form.

Notice that $\mathrm{imad}_m\, N_2 \subset \pi_m\,\mathrm{ad}(j_{m-1}H)(P_H^{(m-1)})$. This allows us to put H into H_2-normal form. The following lemma characterizes a complement to $\mathrm{imad}_m(N_2)$ in P_m and thus also to $\mathrm{imad}_m(N_2) \cap \mathrm{kerad}_m(S_2)$ in $\mathrm{kerad}_m(S_2)$.

2.21. LEMMA. There exists a $Y \in \mathrm{kerad}(S_2)$ such that
$P_m = \mathrm{kerad}_m(Y) \oplus \mathrm{imad}_m(N_2)$.

proof. $N_2 \in P_2$ and P_2 is a semisimple Lie algebra because P_2 is isomorphic to $sp(n,\mathbb{R})$. Thus $\mathrm{kerad}(S_2)$ is a reductive subalgebra of P_2. By the Jacobson-Morosov lemma we may embed N_2 in a Lie subalgebra of $\mathrm{kerad}(S_2)$ isomorphic to $sl(2,\mathbb{R})$. (see Jacobson [1962]), that is, there are elements Y and Z in $\mathrm{kerad}_2(S_2)$ such that

(2.22) $\{N_2,Y\} = Z,\ \{N_2,Z\} = 2N_2,\ \{Y,Z\} = -2Y.$

By composing the embedding of $sl(2,\mathbb{R})$ in $\mathrm{kerad}_2(S)$ with the adjoint representation on P_{m+1} we may regard P_{m+1} as an $sl(2,\mathbb{R})$ module. Hence there is a decomposition of P_{m+1} into irreducible $sl(2,\mathbb{R})$ modules V_i. On each V_i the complement to $\mathrm{imad}_{m+1}(N_2)$ is given by the lowest weight-

space which is just $\mathrm{kerad}_{m+1}(Y)$. Therefore $P_{m+1} = \mathrm{kerad}_{m+1}(Y) \oplus$
$\oplus\ \mathrm{imad}_{m+1}(N_2)$. ⊠

2.23. THEOREM. Let $F \in P$ be in S_2-normal form. Then for each $k \in \mathbb{N}$, $k \geq 2$ there is a $\varphi \in G$ such that $H = F\circ\varphi$ is in H_2-normal form up to order k.

proof. Lemma 2.21. shows that $\tilde{N}(H_2) = \mathrm{kerad}(S) \cap \mathrm{kerad}(Y)$, and by the previous remarks one may by a sequence of transformations transform $j_k H$ into $H_2 + \tilde{N}(H_2)$. ⊠

Notice that also H_2-normal forms are not unique. Even the $sl(2,\mathbb{R})$ embedding need not be unique. At each level one might choose a different Y. Suppose H is in H_2-normal form up to order k then for $3 \leq m \leq k$ $H_m \in \mathrm{kerad}(S_2) \cap \mathrm{kerad}(Y)$ for some Y. Consider $A.H$ where $A.H_2 = H_2$, that is, $A.S = S$, $A.N = N$. Then $\tilde{H}_m = A.H_m \in \mathrm{kerad}(S_2) \cap \mathrm{kerad}(A.Y)$. Thus $\tilde{H}$ is in H_2-normal form but for some different embedding. Of course one can transform $\tilde{H}_m$ back to $\mathrm{kerad}(S_2) \cap \mathrm{kerad}(Y)$ but then one expects to obtain a normal form different from $\tilde{H}$.

3. Examples of normal form computations

3.1. BASIC COMPUTATION. Consider a Hamiltonian system $(\mathbb{R}^{2n},\omega,H)$ with

$$(2.24) \qquad H_2(x,y) = \sum_{j=1}^{n} \tfrac{1}{2}\lambda_j(x_j^2+y_j^2) \qquad ;\lambda_j \in \mathbb{R}$$

Then the matrix corresponding to X_{H_2} has purely imaginary eigenvalues $\pm i\lambda_j$ and is in normal form as an infinitesimal symplectic linear map. It is easy to check that H_2 is semisimple. If we introduce complex conjugate variables $z_j = x_j + iy_j$, $\zeta_j = \bar{z}_j$; $1 \leq j \leq n$, then $\mathrm{ad}(H(x,y))$ becomes

$$(2.25) \qquad \mathrm{ad}(H(z,\zeta)) = -i \sum_{j=1}^{n} \lambda_j\left(z_j \frac{\partial}{\partial z_j} - \zeta_j \frac{\partial}{\partial \zeta_j}\right)$$

Introduce the notation $z^k\zeta^l$ for the monomial $z_1^{k_1}z_2^{k_2}\dots z_n^{k_n}\zeta_1^{l_1}\zeta_2^{l_2}\dots\zeta_n^{l_n}$ where $k = (k_1,\dots,k_n)$, $l = (l_1,\dots,l_n)$ and $k_j, l_j \in \mathbb{N} \cup \{0\}$. Furthermore let $|k| = k_1 +\dots+ k_n$ and $|l| = l_1 +\dots+ l_n$. If $|k| + |l| = m$ then $z^k\zeta^l$ is a monomial of order m.

Now ad(H(z,ζ)) acts diagonally on the space of homogeneous polynomials of order m if we take the monomials $z^k\zeta^l$ as basis.The eigenvalues of ad(H(z,ζ)) are

$$(2.26) \qquad -i \langle\lambda, k - l\rangle = -i \sum_{j=1}^{n} \lambda_j(k_j-l_j)$$

with corresponding eigenvectors $z^k\zeta^l$.

Because H_2 is semisimple we have $N(H_2) = \text{kerad}(H_2)$. It is now obvious that $\text{kerad}(H_2)$ is spanned by those monomials for which $\langle\lambda,k - l\rangle = 0$.

3.2. <u>NONRESONANT CASE</u>. If the λ_j are independent over $\mathbb{Q}$ we speak of a nonresonant system. In this case the only monomials $z^k\zeta^l$ for which $\langle\lambda,k - l\rangle = 0$ are those where $k = l$. Because $z_j\zeta_j = x_j^2 + y_j^2$ it follows immediately that $\text{kerad}(H_2)$ is generated by the homogeneous polynomials $x_j^2 + y_j^2$, $1 \leq j \leq n$. Thus all elements in $\text{kerad}(H_2)$ are of even order. This is just the case considered by Birkhoff [1927] ch.III.8.

The linear symplectic maps which map H_2 to itself form the torus group on $\mathbb{R}^{2n}$ with generators $\text{expad}(x_j^2+y_j^2)$, $1 \leq j \leq n$. By theorem 2.12. all symplectic diffeomorphisms stabilizing the space of normal forms for H are of the form $A\circ\text{expad}(P)$ with A in the above defined torus group and $P \in \text{kerad}(H_2) \cap P_3^+$. Because $\text{kerad}(H_2)$ is an abelian Lie algebra, we may write these stabilizing maps as expad(P), $P \in \text{kerad}(H_2)$. It now easily follows that the Birkhoff normal form in the nonresonant case is unique. (cf. Birkhoff [1927] and the remark of Moser [1968] page 9 and 13).

3.3. <u>RESONANT CASE IN TWO DEGREES OF FREEDOM</u>. When the λ_j are dependent over $\mathbb{Q}$ we speak of a resonant system. If we consider the case of two degrees of freedom, that is, $n = 2$, then $H_2(x,y) = \frac{1}{2}\lambda_1(x_1^2+y_2^2) + \frac{1}{2}\lambda_2(x_2^2+y_2^2)$ with $\lambda_1/\lambda_2 \in \mathbb{Q}$. Without loss of generality we may take $\lambda_1 \in \mathbb{N}$, $\lambda_2 \in \mathbb{Z}$ with g.c.d. $(\lambda_1,\lambda_2) = 1$. Suppose $1 \leqslant \lambda_1 < |\lambda_2|$ (that is, the 1 : 1 resonances are excluded). Then (2.26) becomes $\lambda_1(k_1-l_1) + \lambda_2(k_2-l_2) = 0$. Thus among the generators for kerad(H_2) are those given in the nonresonance case. We call those the Birkhoff generators. In addition to the two Birkhoff generators we find two resonance generators, which are given by $(k_1,k_2,l_1,l_2) = (\lambda_2,0,0,\lambda_1)$ and its complex conjugate $(0,\lambda_1,\lambda_2,0)$ if $\lambda_2 > 0$, and $(|\lambda_2|,\lambda_1,0,0)$ and $(0,0,|\lambda_2|,\lambda_1)$ if $\lambda_2 < 0$. In both the definite $\lambda_2 > 0$ case and the indefinite $\lambda_2 < 0$ case, the above four generators are the generators of kerad(H_2). Therefore we may consider kerad(H_2) as an algebra of formal power series in four variables B_1,B_2,R_1,R_2, where B_1 and B_2 stand for Birkhoff generators (which are quadratic functions) while R_1 and R_2 stand for the resonance generators (which are functions of order $\lambda_1 + |\lambda_2|$). Writing $\lambda_1 = p$ and $\lambda_2 = q$ in the definite case we have:

$$
(2.27^a)\qquad
\begin{aligned}
B_1(x,y) &= H_2(x,y) = \tfrac{1}{2}p(x_1^2+y_1^2) + \tfrac{1}{2}q(x_2^2+y_2^2)\\
B_2(x,y) &= \tfrac{1}{2}p(x_1^2+y_1^2) - \tfrac{1}{2}q(x_2^2+y_2^2)\\
R_1(x,y) &= \tfrac{1}{2}(z_1^q\zeta_2^p+\zeta_1^q z_2^p)\\
R_2(x,y) &= \tfrac{1}{2i}(z_1^q\zeta_2^p-\zeta_1^q z_2^p)
\end{aligned}
$$

with the relation

$$
(2.28)\qquad R_1^2 + R_2^2 = \left(\frac{B_1+B_2}{p}\right)^q\left(\frac{B_1-B_2}{q}\right)^p
$$

With $\lambda_1 = p$ and $|\lambda_2| = q$ in the indefinite case we have:

$$(2.27^b)\quad \begin{aligned} B_1(x,y) &= H_2(x,y) = \tfrac{1}{2}p(x_1^2+y_1^2) - \tfrac{1}{2}q(x_2^2+y_2^2) \\ B_2(x,y) &= \tfrac{1}{2}p(x_1^2+y_1^2) + \tfrac{1}{2}q(x_2^2+y_2^2) \\ R_1(x,y) &= \tfrac{1}{2}(z_1^q z_2^p + \zeta_1^q \zeta_2^p) \\ R_2(x,y) &= \tfrac{1}{2i}(z_1^q z_2^p - \zeta_1^q \zeta_2^p) \end{aligned}$$

with the same relation (2.28). Furthermore in both cases we have the bracket relations

$$(2.29)\quad \begin{aligned} \{B_2,R_1\} &= 2pqR_2 \\ \{B_2,R_2\} &= -2pqR_1 \end{aligned}$$

The action of expad(P) with $P \in \mathrm{kerad}(H_2)$ on a function $F \in \mathrm{kerad}(H_2)$ is now completely determined by the bracket relations among the generators B_1, B_2, R_1, R_2. The subgroup of linear symplectic maps which leave H_2 fixed is generated by expad(B_1) and expad(B_2).

Consider the formal power series case, that is, $\mathrm{kerad}(H_2) = \mathbb{R}[[B_1,B_2,R_1,R_2]]$ and suppose that H is in H_2-normal form. Then $H \in \mathbb{R}[[B_1,B_2,R_1,R_2]]$. Furthermore H has 'linear term'

$$(2.30)\quad H_2 + a_1R_1 + a_2R_2.$$

Suppose that $a_1 \neq 0$ and $a_2 \neq 0$. Clearly $\mathrm{expad}(B_1).H = H$. We now describe the effect of expad(λB_2) on the linear term (2.30). We have

$$\begin{aligned} &\mathrm{expad}(\lambda B_2).(H_2 + a_1R_1 + a_2R_2) = \\ &= H_2 + a_1\ \mathrm{expad}(\lambda B_2).R_1 + a_2\ \mathrm{expad}(\lambda B_2).R_2 = \\ &= H_2 + a_1 \sum_{k=1}^{\infty} \frac{\lambda^k}{k!}\mathrm{ad}^k(B_2)R_1 + a_2 \sum_{k=1}^{\infty} \frac{\lambda^k}{k!}\mathrm{ad}^k(B_2)R_2 \end{aligned}$$

By (2.29) we have $\mathrm{ad}^{2n}(B_2)R_1 = (-1)^n(2pq)^{2n}R_1$ and $\mathrm{ad}^{2n+1}(B_2)R_1 = (-1)^n(2pq)^{2n+1}R_2$, while $\mathrm{ad}^{2n}(B_2)R_2 = (-1)^n(2pq)^{2n}R_2$ and $\mathrm{ad}^{2n+1}(B_2)R_2 = (-1)^{n+1}(2pq)^{2n+1}R_1$. Therefore

$$\exp\mathrm{ad}(\lambda B_2)(H_2 + a_1R_1 + a_2R_2) =$$
$$= H_2 + a_1\cos(2pq\lambda)R_1 + a_1\sin(2pq\lambda)R_2 + a_2\cos(2pq\lambda)R_2 -$$
$$- a_2\sin(2pq\lambda)R_1 =$$
$$= H_2 + (a_1\cos(2pq\lambda) - a_2\sin(2pq\lambda))R_1 + (a_1\sin(2pq\lambda) +$$
$$+ a_2\cos(2pq\lambda))R_2$$

If we consider the $\mathbb{R}$-module with basis B_1,B_2,R_1,R_2 then the action of $\exp\mathrm{ad}(\lambda B_2)$ is nothing more then a rotation in the R_1,R_2-plane over an angle depending on λ. Hence we may choose λ so that $\exp\mathrm{ad}(\lambda B_2)(H_2+a_1R_1+a_2R_2) = H_2 + a_3R_1$. Thus with the appropriate choice of λ, $a_3 = a_1 \cos(2pq\lambda) - a_2 \sin(2pq\lambda)$. We have proved

2.31. <u>LEMMA</u>. If $H \in \mathbb{R}[[B_1,B_2,R_1,R_2]]$ has linear term $L = H_2 + a_1R_1 + a_2R_2$, with at least one of the two coefficients a_1,a_2 nonzero then choosing $\lambda \in \mathbb{R}$ so that $a_1\sin(2pq\lambda) + a_2\cos(2pq\lambda) = 0$ implies that $\exp\mathrm{ad}(\lambda B_2)L = H_2 + a_3R_1$ with $a_3 = a_1\cos(2pq\lambda) - a_2\sin(2pq\lambda)$.

Using the relation (2.28) we may write $\mathbb{R}[[B_1,B_2,R_1,R_2]] = R_2.\mathbb{R}[[B_1,B_2,R_1]] + \mathbb{R}[[B_1,B_2,R_1]]$. If H is in H_2-normal form then $H = H_2 + aR_1 + P_1 + R_2P_2$ with $P_1 \in \mathbb{R}[[B_1,B_2,R_1]]$ of degree ≥ 2 and $P_2 \in \mathbb{R}[[B_1,B_2,R_1]]$ of degree ≥ 1. We will show that by an appropriate transformation we may get rid of the terms in $R_2.\mathbb{R}[[B_1,B_2,R_1]]$. Towards this end we consider a transformation expad(P) with P of order $m > 2$ in the (x,y) variables. On H this transformation has the effect that terms of order $m + p + q-2$ in $\mathrm{imad}(R_1)$ are added to H (recall that R_1 was of order $p + q$ in (x,y) variables). Just as in the normal form theorem 2.5., it follows that we may find a transformation which eliminates all terms in $\mathrm{imad}(R_1)$.

2.32. LEMMA. All elements of $R_2.\mathbb{R}[[B_1,B_2,R_1]]$ are in $\mathrm{imad}(R_1)$.

proof. Consider $R_2B_1^{\alpha}B_2^{\gamma}R_1^{\delta}$ then $\mathrm{ad}(R_1)(-\frac{1}{2pq(\gamma+1)}B_1^{\alpha}B_2^{\gamma+1}R_1^{\delta}) =$
$= R_2B_1^{\alpha}B_2^{\gamma}R_1^{\delta}$ by (2.29). ⊠

2.33. THEOREM. Suppose H is the Hamiltonian of a two degree of freedom system in resonance. Furthermore suppose that H is in H_2-normal form and fulfills the condition of lemma 2.31. Then there exists a symplectic formal power series transformation φ such that $H\circ\varphi \in \mathbb{R}[[B_1,B_2,R_1]]$.

proof. Follows from lemmas 2.31 and 2.32. ⊠

2.34. REMARK. Of course the above also holds if we replace R_1 by R_2.

2.35. REMARK. Theorem 2.33. in fact shows part of the normalization of H with respect to $H_2 + a_3R_1$. For a complete $H_2 + a_3R_1$-normal form for H we have to find an appropriate complement to $\mathrm{imad}(R_1)$. In general such a complement is not easy to find.

2.36. REMARK. In lemma 2.31. we have used all the freedom of choice for the linear part of our transformations mapping normal forms to normal forms. In this case this means that the constant m_0 of proposition 2.17 is equal to $p + q$, $p + q$ being the degree of R_1 in (x,y) coordinates.

3.4. RESONANT CASE IN n DEGREES OF FREEDOM. For resonant n degree of freedom systems we only make some general remarks. Let H_2 be as in (2.24) then H_2 is semisimple and consequently $j_m(\mathrm{kerad}(H_2))$ is a reductive subalgebra of P_m^- for all m. Thus we may write $j_m(\mathrm{kerad}(H_2)) = C + j_m\{j_m(\mathrm{kerad}(H_2)), j_m(\mathrm{kerad}(H_2))\}$ where C is the center of $j_m(\mathrm{kerad}(H_2))$. If H is in H_2-normal form and $j_mH \in C$ then it is obvious that j_mH is unique by theorem 2.12. It is well known that among the generators of $\mathrm{kerad}(H_2)$ there are n Birkhoff generators. Furthermore up to the order where the first resonance generators appear H is a function of these Birkhoff generators alone

2.37. LEMMA. If H is in H_2-normal form with H_2 as in (2.24) then H is uniquely determined up to degree $m - 1$ if m is the minimal order where resonance terms appear.

At the order where the resonance terms appear for the first time we may use the linear group of transformations which map H_2 to itself to further normalize the resonance terms. This is illustrated in Duistermaat [1983[a]] for the 1 : 1 : 2 resonance and in Dell'Antonio et al. for the 1 : 1 resonances.

Finally we will say something about the case of simple resonance, that is, there is just one dependence relation (over $\mathbb{Q}$) among the eigenvalues λ_i. If this dependence relation involves only two of the eigenvalues then the normal form computations reduce to those of two degrees of freedom because only the variables corresponding to these two λ_j's are involved. If the dependence relation involves more eigenvalues then the computations become more complicated but one may proceed along the same lines as in the case of two degrees of freedom.

3.5. THE NONSEMISIMPLE 1 : -1 RESONANCE. In this last part of section 3 we will treat the normalization of the Hamiltonian of a system of two degrees of freedom in nonsemisimple 1 : -1 resonance. This is the case which is the topic of the next chapters. The quadratic part of the Hamiltonian (which differs in an essential way from the one given by (2.24)) is given by

$$(2.38) \qquad H_2(x,y) = \alpha(x_1y_2-x_2y_1) + \tfrac{1}{2}(x_1^2+x_2^2)$$

As is easily checked the semisimple part of H_2 is $\alpha(x_1y_2-x_2y_1)$, which is in 1 : -1 resonance. Furthermore H_2 has a nonzero nilpotent part $\frac{1}{2}(x_1^2+x_2^2)$. For convenience take $\alpha = 1$. Then the semisimple part of H_2 is $S(x,y) = x_1y_2 - x_2y_1$ and the nilpotent part is $X(x,y) = \frac{1}{2}(x_1^2+x_2^2)$; furthermore $\{X,S\} = 0$.

The generators of kerad(S) are again found by introducing complex conjugate variables. However, because the normal form for S given above is different from the one given by (2.24), the complex conjugate coordinates diagonalizing ad(S) are different from the ones given in subsection 3.1. In this case we take $z_1 = x_1 + ix_2$, $z_2 = y_1 + iy_2$, $\zeta_1 = \bar{z}_1$ and $\zeta_2 = \bar{z}_2$. Then

$$\text{(2.39)} \qquad \operatorname{ad}(S(z,\zeta)) = i \sum_{j=1}^{2} \left(z_j \frac{\partial}{\partial z_j} - \zeta_j \frac{\partial}{\partial \zeta_j}\right),$$

kerad(S) is generated by $S(x,y)$, $X(x,y)$, $Y(x,y) = \frac{1}{2}(y_1^2+y_2^2)$ and $Z(x,y) = x_1y_1 + x_2y_2$, which satisfy the relation

$$\text{(2.40)} \qquad S^2 + Z^2 = 4XY, \; X \geqslant 0, \; Y \geqslant 0.$$

Furthermore $\{Y,X\} = Z$, $\{Z,X\} = 2X$, $\{Z,Y\} = -2Y$, that is X,Y,Z span a Lie subalgebra of $\text{kerad}_2(S)$ isomorphic to $sl(2,\mathbb{R})$. Hence the choice of an embedding of X in a Lie algebra isomorphic to $sl(2,\mathbb{R})$ is immediately clear. According to lemma 2.21., $\tilde{N}(H_2) = \text{kerad}(S) \cap \text{kerad}(Y)$. It is easy to check that $\tilde{N}(H_2) = \mathbb{R}[[S,Y]]$ when written as a space of formal power series.

Any map in G which leaves S fixed is a Lie algebra isomorphism for the $sl(2,\mathbb{R})$ spanned by X,Y,Z. However the above choice of the span of $sl(2,\mathbb{R})$ is the most simple one. Any non-identity Lie algebra isomorphism will give rise to more complicated functions and thus to a more complicated description of $N(H_2)$. Let φ be a mapping, which takes an H_2-normal form to an H_2-normal form. If we require that φ leaves Y fixed, then one may easily check that $\varphi = A \circ \exp ad(P)$ with $P \in P_3^+ \cap \ker ad(S) \cap \ker ad(X) \cap \ker ad(Y)$ and $A.S = S$, $A.X = X$, $A.Y = Y$. Thus φ is of the form $\exp ad(P)$ with P a polynomial in S only. But then φ acts as the identity on an H_2-normal form.

2.41. LEMMA. Suppose H is in H_2-normal form with H_2 given by (2.38). Furthermore suppose that the $sl(2,\mathbb{R})$ embedding of X is fixed throughout. Then the H_2-normal form is unique.

4. Integrals and energy-momentum maps

Suppose we have a formal Hamiltonian system $(\mathbb{R}^{2n},\omega,H)$ with H in S_H-normal form, where S_H is the semisimple part of H_2, $S_H \neq 0$. Let F be an integral for the system $(\mathbb{R}^{2n},\omega,H)$.

2.42. THEOREM. If F is an integral for $(\mathbb{R}^{2n},\omega,H)$ with H in S_H-normal form then $F \in \ker ad(S_H)$.

proof. The proof is straight forward using induction. We have $0 = \pi_2\{H,F\} = \{H_2,F_2\}$. Thus $F_2 \in \ker ad(S_H)$. Also $0 = \pi_3\{H,F\} = \{H_2,F_3\} + \{H_3,F_2\}$. Obviously $\{H_3,F_2\} \in \ker ad(S_H)$ and thus $\{H_2,F_3\} \in \ker ad(S_H)$. Therefore by lemma 2.6. we have $F_3 \in \ker ad(S_H)$. Now suppose $j_kF \in \ker ad(S_H)$. Then $\pi_{k+1}\{H,F\} = \{H_2,F_{k+1}\}$ + (brackets of elements in $\ker ad(S_H)$). As before it follows that $F_{k+1} \in \ker ad(S_H)$, which proves the induction step. ⊠

Thus $F \in \ker ad(S_H)$. We may now normalize F with respect to F_2 within $\ker ad(S_H)$. Such a normalizing transformation maps H to some function $\tilde{H}$ which again is in S_H-normal form. If the semisimple part S_F of F_2 is nonzero. Then putting F in S_F-normal form within $\ker ad(S_H)$ leads to the following.

2.43. THEOREM. Suppose F is an integral for the Hamiltonian system $(\mathbb{R}^{2n},\omega,H)$. Furthermore suppose that S_H and S_F are nonzero. Then there exists a symplectic formal power series transformation φ such that $\varphi.H \in \mathrm{kerad}(S_F) \cap \mathrm{kerad}(S_H)$ and $\varphi.F \in \mathrm{kerad}(S_F) \cap \mathrm{kerad}(S_H)$.

Restricting to systems with two degrees of freedom, theorem 2.43. has some direct consequences, which we now discuss. Consider a system $(\mathbb{R}^4,\omega,H)$ with integral F. Furthermore suppose that S_H and S_F are nonzero and independent. Let $H \times F$ be the energy-momentum map of $(\mathbb{R}^4,\omega,H)$. For $\varphi \in G$ define an action of G on the energy-momentum map by $\varphi.(H\times F) = (H\circ\varphi) \times (F\circ\varphi)$. Taking the normalizing transformation given by theorem 2.43. we obtain an energy-momentum map $(H\circ\varphi) \times (F\circ\varphi)$ where $H\circ\varphi = \tilde{H}$ and $F\circ\varphi = \tilde{F}$ are formal power series in S_H and S_F because S_F and S_H generate a maximal abelian subalgebra and thus $\mathrm{kerad}(S_F) \cap \mathrm{kerad}(S_H)$ is generated by S_H and S_F only. Applying a formal power series diffeomorphism on the target space $\mathbb{R}^2$ of $\tilde{H} \times \tilde{F}$ reduces $\tilde{H} \times \tilde{F}$ to $S_H \times S_F$. Thus working with formal power series a right-left action reduces $H \times F$ to $S_H \times S_F$ (cf. Eliasson [1984] and the theory of the next chapter).

5. Historical notes

It is well known that mathematicians often try to simplify the formulas in a problem in order to make it easier to find a solution. Such a simplified formula is then called a "normal form". From this point of view a normal form is nothing more than a relatively simple form for a formula (which for instance describes a map, function or differential equation) which serves the purpose of the user best.

One of the best known examples of the development of a concept of normalization of course is the theory of normal forms for matrices. This theory has a close relation to the theory of normal forms for linear differential equations or, in the Hamiltonian case, to the normal forms for homogeneous quadratic Hamiltonian functions (see Burgoyne and Cushman [1974, 1976], Williamson [1936] and chapter 1).

For normal forms of ordinary differential equations one might go back to a memoir of Briot and Bouquet [1854], although most people give credit to Poincaré [1879] who made several important contributions. Other important contributions were made by Liapunov [1892], Dulac [1912], Siegel [1952] and Sternberg [1958,1959,1961]. In Kelley [1963] one finds a discussion of the developments until 1963. More recent contributions are those of Brjuno [1971,1972], Takens [1974] and Broer [1979,1980]. Most of these papers treat the general case of ordinary differential equations. As with Hamiltonians there are different cases to consider according to the configuration of the eigenvalues of the linearized system.

Sternberg [1958,1959,1961] extensively treats problems concerning normalization and classification of vector fields. He uses the concepts of formal power series, Lie groups and Lie algebras. Based on Sternberg's work, Chen [1963] formulates some formal theorems (section 8 of the paper) which very much resemble our approach. The aim of Chen's paper was to find linear normal forms for systems having eigenvalues with nonzero real part. A theorem similar to our theorem 2.5. can be found in Takens [1974] who also refers to Sternberg as the source of the basic ideas. In Brjuno [1971,1972] one finds a extensive treatment of normal forms for vector fields which treats all cases and deals with the Hamiltonian case seperately. His work is mostly based on power series methods. He also poses the question of uniqueness and the existence of invariants. In fact he proves the only if part of our theorem 2.7. Broer's work is based on Takens' theorem. His approach of normalizing jets of vector fields is parallel to our formal approach.

If we consider papers which treat only nonlinear Hamiltonian systems, we may go back to Whittaker [1901] (also Whittaker [1917] ch. XVI) who simplifies Hamilton's equations using canonical changes of coordinates. His method, which was inspired by Delaunay's lunar theory, makes use of trigonometric series. At present the main reference for normal forms for Hamiltonian systems is Birkhoff [1927]. Although not explicitly stated, he uses the concept of Poisson bracket. Probably because of his restricted purposes Birkhoff only states the theory for the case in which the linearized system has nonresonant purely imaginary eigenvalues. His theory is in terms of formal power series, as is ours. In the same year Cherry [1927] published closely related results.

Also Siegel's [1952,1956] work on normal forms for Hamiltonian systems has to be mentioned. His paper mostly deals with convergence problems of the series involved. Siegel [1956] goes into the problem of uniqueness for normal forms for area preserving mappings.

In 1958 Moser stated a normal form theorem in a setting of formal power series, which was in fact an extension of the Birkhoff approach to the resonant case. This was taken up by Gustavson [1966] in his paper on formal integrals. Since then normal forms for Hamiltonian systems in resonance are often referred to as Gustavson-normal forms. Gustavson's approach in just a constructive treatment of the formal statements of Chen [1963]. (compare our subsections 3.3 and 3.4). In Moser [1968] one can find a nice resumé of the theory of formal normal forms for Hamiltonian systems. He also makes some remarks on the uniqueness of Hamiltonian normal forms (the Birkhoff case) and considers uniqueness of normal forms for area preserving maps.

Since 1968 a number of papers have appeared which deal with normal forms, normalizing transformations and integrals in terms of formal power series. For instance Deprit [1969] treats transformations of formal power series, Deprit et al [1969] and Meyer [1974^b] deal with normal form algorithms and Giorgilli and Galgani [1978] are concerned with algorithms for finding formal integrals related to normalizing transformations.

In Meyer and Schmidt [1971] one finds for the first time a normal form for a Hamiltonian function whose quadratic terms are not semisimple. Such a nonsemisimple case arises in the restricted problem of three bodies. A real variable method obtaining a normal form in this case is given in van der Meer [1982]. In Cushman et al. [1983] this method is extended using representation theory (compare our theorem 2.21 and 2.23). Also Deprit [1983] gives a normalizing algorithm for the nonsemisimple case, based on the two preceding ones, but more fit for use on a computer.

Other recent papers dealing with normal forms for Hamiltonian functions in classical mechanical systems are Deprit [1981,1982] which eliminates terms in power series in Delaunay variables, and Cushman [1984] which considers normalization of power series in ε with smooth coefficients with respect to a term of order zero whose corresponding vector field has only periodic orbits.

The problem of uniqueness of normal forms of Hamiltonian functions treated in this chapter occurs in Brjuno [1971]. The area preserving map case is treated in Siegel [1956] and Moser [1968]. For the nonsemisimple 1 : -1 resonance problem nonuniqueness became apparent in van der Meer [1982] because of the freedom in choosing the complement of $\mathrm{imad}(H_2)$.

Chapter III

Fibration preserving normal forms for energy-momentum maps

0. Introduction

Consider a Hamiltonian system $(\mathbb{R}^4,\omega,H)$, $H \in C^\infty(\mathbb{R}^4,\mathbb{R})$. Let S be the semisimple part of H_2 and suppose that H is in S-normal form. Thus the system is integrable with integral S. Furthermore suppose that the action of X_S is a circle action. If we consider the action of the group of symplectic C^∞-diffeomorphisms commuting with the action of X_S on H, then the ∞-jet of H will have infinitely many non zero terms. This means that we cannot transform H to a finite part of its Taylor expansion by means of such a symplectic transformation. If instead of the function H we look at the energy-momentum map $H \times S$ then we meet the same problem using the right-action (that is, on the source) of the symplectic C^∞-diffeomorphisms equivariant with respect to the action of X_S. If we drop the restriction that the diffeomorphisms be symplectic and consider a right-left action then $H \times S$ can be drastically simplified. Because $H \times S$ is invariant with respect to the S^1-action of X_S we are in the framework of stability of invariant (or equivariant) maps.

In this chapter we will show how the theory of stability of maps in the equivariant version by Poenaru [1976] and Bierstone [1980] (also see Izumiya [1982] and Roberts [1983]) can be used to obtain normal forms for energy-momentum maps of Hamiltonian systems with S^1-symmetry. It turns out that the resulting normal form again can be considered as an energy-momentum map for a Hamiltonian system. The fibration of the normal form and the original energy-momentum map are the same up to diffeomorphism, that is, the normalization is fibration preserving. Therefore one can obtain qualitative information about the original system from the system corresponding to the normalized map.

In the first section we state the basic facts from the theory of stability of equivariant maps. In section 2 we show how these facts can be applied to energy-momentum maps and what the consequences are for the qualitative behaviour of the related Hamiltonian systems. In section 3 we apply the theory to the concrete example of the Hamiltonian Hopf bifurcation. In the final section we discuss the results.

The theory of stability of functions and maps has been developed by Mather [1968-1970] (see Martinet [1982] for a nice treatment). Its extension to the case of equivariant real valued C^∞-functions is due to Poenaru [1976] and Wassermann [1977]; while the extension to equivariant C^∞-maps can be found in Poenaru [1976] and Bierstone [1980] (also see Roberts [1983] and Izumiya [1982]).

1. Preliminaries from the theory of stability of maps

In this section we state without proof some facts from the theory of stability of maps which will be used in the following. The proofs and further background can be found in the literature cited above.

Consider the space $C^\infty(\mathbb{R}^n)$ of smooth real valued functions on $\mathbb{R}^n$. If $x_1,\dots,x_n$ are the coordinates on $\mathbb{R}^n$ then $\mathbb{R}[x] = \mathbb{R}[x_1,\dots,x_n]$ is the subspace of polynomials on $\mathbb{R}^n$. Let S be a Lie group acting linearly on $\mathbb{R}^n$. Then the S-action induces an action on $C^\infty(\mathbb{R}^n)$ (and also of course on the algebra of polynomials) defined by

(3.1) $\varphi \,.\, F(a) = F(\varphi^{-1}(x)),\ \varphi \in S,\ F \in C^\infty(\mathbb{R}^n).$

3.2. DEFINITION. A function $F \in C^\infty(\mathbb{R}^n)$ is S-invariant if $\varphi\,.F = F$ for all $\varphi \in S$.

This is equivalent with saying $F = F \circ \varphi$ for all $\varphi \in S$. Denote the space of smooth S-invariant functions by $C^\infty(\mathbb{R}^n)^S$ and similarly the algebra of S-invariant polynomials by $\mathbb{R}[x]^S$. We have the following theorem due to Hilbert (see Poenaru [1976] for a proof).

3.3. THEOREM. Let S be a compact Lie group acting linearly on $\mathbb{R}^n$. Then there exist finitely many invariant polynomials $\rho_1(x),\dots,\rho_k(x) \in \mathbb{R}[x]^S$ which generate $\mathbb{R}[x]^S$ as an $\mathbb{R}$-algebra.

An alternative formulation of this theorem is: there exists an algebraic map $\rho\colon \mathbb{R}^n \to \mathbb{R}^k$ defined by $\rho(x) = (\rho_1(x),\dots,\rho_k(x)) = y$ such that the pull-back $\mathbb{R}[x]^S \xleftarrow{\rho^*} \mathbb{R}[y]$ is surjective. We call the polynomials ρ_i the *Hilbert generators* for $\mathbb{R}[x]^S$ and the mapping ρ a *Hilbert map* for the action of S. Notice that without loss of generality one may always take the ρ_i to be homogeneous of degree greater than zero.

Schwarz [1975] proved an important extension of Hilbert's theorem to the case of C^∞-functions.

3.4. THEOREM. (Schwarz [1975]) If ρ is a Hilbert map for the S-action then the map $C^\infty(\mathbb{R}^n)^S \xleftarrow{\rho^*} C^\infty(\mathbb{R}^k)$ is surjective.

This means that every S-invariant smooth function can be written as a smooth function in the finitely many Hilbert generators of $\mathbb{R}[x]^S$. Theorem 3.4. together with the equivariant preparation theorem (see Poenaru [1976]) form the basis of the theory of equivariant stability of equivariant maps. We will now state the main results of this theory for the special case of invariant maps.

Let $C^\infty(\mathbb{R}^n,\mathbb{R}^p)$ be the space of C^∞-maps $F\colon \mathbb{R}^n \to \mathbb{R}^p$. Furthermore let S be a compact group acting linearly on $\mathbb{R}^n$ as well as $\mathbb{R}^p$.

3.5. DEFINITION. A map $F \in C^\infty(\mathbb{R}^n,\mathbb{R}^p)$ is *S-equivariant* if $\varphi \,.\, F(x) = F(\varphi x)$ for all $\varphi \in S$. In the left hand side we consider the action of S on $\mathbb{R}^p$, in the right hand side the action of S on $\mathbb{R}^n$.

If S acts trivially on $\mathbb{R}^p$, that is, if $\varphi \,.\, x = x$ for all $x \in \mathbb{R}^p$,

$\varphi \in S$, then definition 3.5. turns into the $C^\infty(\mathbb{R}^n,\mathbb{R}^p)$ analogue of definition 3.2. and the theory for S-equivariant maps naturally restricts to a theory for S-invariant maps. We shall apply the equivariant theory with the trivial action of S on $\mathbb{R}^p$.

Let $C^\infty(\mathbb{R}^n,\mathbb{R}^p)^S$ denote the subspace of $C^\infty(\mathbb{R}^n,\mathbb{R}^p)$ of S-invariant maps. Furthermore let $\mathrm{Diff}_S(\mathbb{R}^n)$ denote the space of S-equivariant diffeomorphisms from $\mathbb{R}^n$ to $\mathbb{R}^n$ and let $\mathrm{Diff}(\mathbb{R}^p)$ denote the space of diffeomorphisms from $\mathbb{R}^p$ to $\mathbb{R}^p$. There is a natural action of $\mathrm{Diff}_S(\mathbb{R}^n) \times \mathrm{Diff}(\mathbb{R}^p)$ on $C^\infty(\mathbb{R}^n,\mathbb{R}^p)^S$ defined by

$$(3.6) \qquad (\varphi,\psi) . F = \psi \circ F \circ \varphi^{-1}.$$

In the following we will consider the local theory of invariant maps, that is, we consider maps $F \in M(n,p)$ where $M(n,p) = \{F \in C^\infty(\mathbb{R}^n,\mathbb{R}^p);\ F(0) = 0\}$, and the group of origin preserving diffeomorphisms on source $\mathbb{R}^n$ and target $\mathbb{R}^p$ which we denote by $\mathrm{Diff}_S(\mathbb{R}^n)_0 \times \mathrm{Diff}(\mathbb{R}^p)_0$. Note that each component of $M(n,p)$ is the maximal ideal $M(n)$, in $C^\infty(\mathbb{R}^n)$. Furthermore let $M(n,p)^S = M(n,p) \cap C^\infty(\mathbb{R}^n,\mathbb{R}^p)^S$.

3.7. <u>DEFINITION</u>. Two elements $F, \tilde{F} \in M(n,p)^S$ are *equivalent* (sometimes called right-left equivalence) if F and $\tilde{F}$ are in the same $\mathrm{Diff}_S(\mathbb{R}^n)_0 \times \mathrm{Diff}(\mathbb{R}^p)_0$ orbit, that is, if there is a $(\varphi,\psi) \in \mathrm{Diff}_S(\mathbb{R}^n) \times \mathrm{Diff}(\mathbb{R}^p)_0$ such that $\tilde{F} = \psi \circ F \circ \varphi^{-1}$.

3.8. <u>DEFINITION</u>. A map $F \in V \subset M(n,p)^S$ is *stable in* V if there is a neighbourhood $U \subset V$ of F (in the appropriate topology) such that each $\tilde{F} \in U$ is equivalent to F. If $V = M(n,p)^S$ then we say that F is *stable*.

In other words, F is stable, if the $\mathrm{Diff}_S(\mathbb{R}^n)_0 \times \mathrm{Diff}(\mathbb{R}^p)_0$ -orbit through F is open. Notice that the notions of equivalence and stability defined above depend on the group action, the group and the space on which the group acts.

For $F \in M(n,p)$, *a smooth vector field along* F is a smooth map ξ of $\mathbb{R}^n$ into the tangent bundle $T\mathbb{R}^p$ of $\mathbb{R}^p$ such that $\pi \circ \xi = F$. Here π denotes the canonical projection of $T\mathbb{R}^p$ to $\mathbb{R}^p$. Let $\Theta(F)$ denote the $C^\infty(\mathbb{R}^n)^S$-module of smooth origin preserving S-equivariant vector fields along F. $\Theta(F)$ can be identified with $M(n,p)^S$. Furthermore let $\Theta(\mathbb{R}^n)$ be the $C^\infty(\mathbb{R}_n)^S$-module of smooth origin preserving S-equivariant vector fields on $\mathbb{R}^n$ and let $\Theta(\mathbb{R}^p)$ be the $C^\infty(\mathbb{R}^p)$-module of origin preserving vector fields on $\mathbb{R}^p$. Define $\alpha_F\colon \Theta(\mathbb{R}^n) \to \Theta(F)$ and $\beta_F\colon \Theta(\mathbb{R}^p) \to \Theta(F)$ by $\alpha_F(\xi) = -dF \circ \xi$ and $\beta_F(\eta) = \eta \circ F$.

3.9. <u>DEFINITION</u>. $F \in M(n,p)^S$ is *infinitesimally stable* if

$$\Theta(F) = \alpha_F(\Theta(\mathbb{R}^n)) + \beta_F(\Theta(\mathbb{R}^p)). \tag{3.10}$$

The right hand side of (3.10) is equal to the tangent space to the orbit of F in the sense that a tangent vector is defined as $\frac{d}{dt}\, \psi_t \circ F \circ \varphi_t^{-1} \Big|_{t=0}$ where ψ_t and φ_t are smooth curves in $\mathrm{Diff}(\mathbb{R}^p)_0$ and $\mathrm{Diff}_S(\mathbb{R}^n)_0$ with $\psi_0 = I$ and $\varphi_0 = I$. The left hand side can be regarded as the tangent space to $M(n,p)^S$ at F. Notice that the definition of $\alpha_F(\Theta(\mathbb{R}^n))$ for $p = 1$ coincides with the definition of S-invariant Jacobian ideal as given in Poenaru [1976] and Wasserman [1977] for the action of $\mathrm{Diff}_S(\mathbb{R}^n)_0$. Therefore we will write $\alpha_F(\Theta(\mathbb{R}^n)) = J(F)^S$ and speak of it as the *Jacobian module*. Let $F_p(F_1,\ldots,F_p)$ denote the space of smooth maps $B\colon \mathbb{R}^n \to \mathbb{R}^p$ such that the component functions $B_i\colon i = 1,\ldots,p$ are smooth functions in the component functions F_i of F. If $F \in M(n,p)^S$ then $B \in M(n,p)^S$ and $F_p(F_1,\ldots,F_p) = \beta_F(\Theta(\mathbb{R}^p)) \subset M(n,p)^S$. Thus we may rephrase the definition of infinitesimal stability as

$$M(n,p)^S = J(F)^S + F_p(F_1,\ldots,F_p). \tag{3.11}$$

3.12. <u>THEOREM</u>. $F \in M(n,p)^S$ is stable (in the sense of definition 3.8.) if and only if F is infinitesimally stable (in the sense of definition 3.9.).

Theorem 3.12. is the main theorem of equivariant singularities theory. The if-part has been proved by Poenaru [1976] and the only if part by Bierstone [1980] (ch.5).

If F is not infinitesimally stable then we may try to determine for which F the codimension of $J(F)^S + F_p(F_1,\ldots,F_p)$ in $M(n,p)^S$ is finite.

3.13. <u>DEFINITION</u>. The *codimension of* $F \in M(n,p)^S$ is

$$\dim \left[M(n,p)^S \Big/ J(F)^S + F_p(F_1,\ldots,F_p) \right].$$

Next consider the Taylor expansion of $F \in M(n,p)^S$. We speak of the $(k_1,\ldots,k_p)$-jet of F if we consider the k_j-jet of F_j for $j = 1,\ldots,p$.

3.14. <u>DEFINITION</u>. A map F is $(k_1,\ldots,k_p)$-*determined* if any map with the same $(k_1,\ldots,k_p)$-jet as F is equivalent to F. A map F is *finitely determined* if there are $k_1,\ldots,k_p$ such that F is $(k_1,\ldots,k_p)$-determined.

We have

3.15. <u>PROPOSITION</u> (cf. Roberts [1983], th. 3.1). $F \in M(n,p)^S$ is finitely determined if and only if F is of finite codimension.

If a map F is $(k_1,\ldots,k_p)$-determined then we may take its $(k_1,\ldots,k_p)$-jet P as a representative for the orbit through F and codim(P) = codim(F). The codimension of P can be determined by considering only finite jets. Codim(P) is just the codimension of $J(P)^S + F_p(P_1,\ldots,P_p) \Big/ (M(n,1)^S)^{k_1} \times \ldots \times (M(n,1)^S)^{k_p}$ in

$M(n,p)^S / (M(n,1)^S)^{k_1} \times \ldots \times (M(n,1)^S)^{k_1}$

If P has nonzero but finite codimension, P is not stable. However one can construct a stable deformation of P depending on finitely many parameters.

3.16. DEFINITION. A map $F_\mu \in M(n,p)^S$ depending smoothly on $\mu = (\mu_1,\ldots,\mu_s)$ is a *deformation* of F if $F_0 = F$.

Let P be the $(k_1,\ldots,k_p)$-jet of a $(k_1,\ldots,k_p)$-determined map F. Choose a basis $b_1,\ldots,b_s$ of the complement of $J(P)^S + F_p(P_1,\ldots,P_p)$ in $M(n,p)^S$ and consider the deformation $P_\mu = P + \mu_1 b_1 + \ldots + \mu_s b_s$. One may consider P_μ as an element of the subspace $M(n+s,p+s)$ of $C^\infty(\mathbb{R}^n \times \mathbb{R}^s, \mathbb{R}^p \times \mathbb{R}^s)$. Extend the action of the compact group S on $\mathbb{R}^n$ to an action of S_μ on $\mathbb{R}^n \times \mathbb{R}^s$ by letting S_μ act trivially on $\mathbb{R}^s$. Now consider the action of $\mathrm{Diff}_S(\mathbb{R}^n \times \mathbb{R}^s)_0 \times \mathrm{Diff}(\mathbb{R}^p \times \mathbb{R}^s)_0$ on $M(n+s,p+s)^S$. We have

$$J(P_\mu)^S + F_{p+s}(P_{\mu,1},\ldots,P_{\mu,p},\mu_1,\ldots,\mu_s) = J(P)^S + \left\langle\frac{\partial P}{\partial \mu_1},\ldots,\frac{\partial P}{\partial \mu_s}\right\rangle$$

$$+ F_{p+s}(P_{\mu,1},\ldots,P_{\mu,p},\mu_1,\ldots,\mu_s) = M(n+s,p+s)^S$$

Thus P_μ is stable in $M(n+s,p+s)^S$. Moreover s is the minimal number of parameters needed to obtain such a stable deformation. The above result is a straight forward extension of the theory of universal deformations of Poenaru [1976] for equivariant right-action to the case of right-left-action. Consequently the deformation P_μ of P is a universal deformation (or universal unfolding) of P.

3.17. DEFINITION. G_ν is a *universal deformation* of G if for any other deformation G_μ of G there exists a smooth map χ between the parameter spaces, $\chi(\mu) = \nu$, $\chi(0) = 0$, such that the pull-back $\chi^* G_\nu$ of G_ν by χ is equivalent to G_μ by means of a μ-dependent transformation $(\varphi,\psi) \in \in \mathrm{Diff}_S(\mathbb{R}^n)_0 \times \mathrm{Diff}(\mathbb{R}^p)_0$, that is, $\psi \circ \chi^* G_\nu \circ \varphi^{-1} = G_\mu$.

3.18. PROPOSITION. Let F_μ be a deformation of F_{μ_0}, with F_{μ_0} $(k_1,\ldots,k_p)$-determined. Let G_ν be the universal deformation of the $(k_1,\ldots,k_p)$-jet G of F_{μ_0}. Then there exists a smooth map χ between the parameter spaces, $\chi(\mu) = \nu$; $\chi(\mu_0) = 0$ and a μ-dependent transformation $(\varphi,\psi) \in \in \mathrm{Diff}_S(\mathbb{R}^n)_0 \times \mathrm{Diff}(\mathbb{R}^p)_0$ such that $\psi \circ \chi^* G_\nu \circ \varphi^{-1} = F_\mu$.

This concludes our statement of results and definitions needed in section 2 and 3.

Note that the complete theory of this section is in terms of S-invariant maps. In the actual computations it is therefore convenient to express all maps as functions or polynomials in the Hilbert generators ρ_i corresponding to the S-action. Consequently one has to know the ρ_i explicitly. This procedure is followed in section 3.

2. Standard forms for energy-momentum maps and invariant sets

In this section we will apply the theory of stability of equivariant maps to the energy-momentum map of a Hamiltonian system having a quadratic integral with periodic flow. Obviously the energy-momentum map is invariant with respect to the flow of the quadratic integral which gives rise to a compact one-parameter group. If the energy-momentum map is finitely determined we need only consider a finite jet of this map. When this finite jet takes a sufficiently simple form we will call it a normal form for the energy-momentum map (notice that this notion of normal form differs from the notion of normal form of a singularity as defined by Arnold [1975]). By construction such a simple finite jet still contains a great deal of qualitative information about the original system, especially when the Hamiltonian system has two degrees of freedom.

Consider a Hamiltonian system $(\mathbb{R}^{2n},\omega,H)$ with ω the standard symplectic form on $\mathbb{R}^{2n}$. Suppose that $H(0) = dH(0) = 0$. The space of smooth functions whose 1-jet at zero vanishes will be denoted $C^\infty_*(\mathbb{R}^{2n},\mathbb{R})$. Suppose furthermore that the system $(\mathbb{R}^{2n},\omega,H)$ posesses an independent quadratic integral $S \in C^\infty_*(\mathbb{R}^{2n},\mathbb{R})$ such that the flow of X_S gives rise to a linear S^1-action on $\mathbb{R}^{2n}$. Let $\mathcal{S}$ be the compact one parameter group generated by the flow of X_S. Furthermore consider the energy-momentum map $H \times S$ defined by $H \times S: z \in \mathbb{R}^{2n} \to (H(z),S(z)) \in \mathbb{R}^2$. Every fiber of the map $H \times S$ is an invariant set for X_H. In particular every fiber is $\mathcal{S}$-invariant.

Next consider the groups $\mathrm{Diff}_{\mathcal{S}}(\mathbb{R}^{2n})_0$ and $\mathrm{Diff}(\mathbb{R}^2)_0$ of origin preserving diffeomorphisms acting on source and target of $H \times S$ respectively. Following the theory of section 1 we may ask whether $H \times S$ is finitely determined with respect to the induced action of $\mathrm{Diff}_{\mathcal{S}}(\mathbb{R}^{2n})_0 \times \mathrm{Diff}(\mathbb{R}^2)_0$ on $C^\infty_*(\mathbb{R}^{2n},\mathbb{R}^2)^{\mathcal{S}}$. When $H \times S$ is finitely determined we may normalize $H \times S$ by a map $(\varphi,\psi) \in \mathrm{Diff}_{\mathcal{S}}(\mathbb{R}^{2n})_0 \times \mathrm{Diff}(\mathbb{R}^2)_0$. $\varphi \in \mathrm{Diff}_{\mathcal{S}}(\mathbb{R}^{2n})_0$ maps a fibre of $H \times S$ to a fibre of $(H \times S) \circ \varphi$, while $\psi \in \mathrm{Diff}(\mathbb{R}^2)_0$ only changes the base points. We have

3.19. <u>THEOREM</u>. Normalizing maps in $C^\infty_*(\mathbb{R}^{2n},\mathbb{R}^2)$ preserves the fibration of $\mathbb{R}^{2n}$ up to a diffeomorphism $\varphi \in \mathrm{Diff}_{\mathcal{S}}(\mathbb{R}^{2n})_0$.

If we consider systems of two degrees of freedom, that is, $H \in C^\infty_*(\mathbb{R}^4,\mathbb{R})$, then the possible fibers of $H \times S$ are easily classified. A regular value of $H \times S$ gives rise to a smooth two dimensional manifold whose connected components must be either a torus or a cylinder due to the S^1-symmetry (see the theory of chapter 4). A singular value of $H \times S$ gives rise to a point, a one dimensional fibre or a critical two dimensional fibre. In the case that the fibre is just a point, we have an equilibrium point for our system. Because the fibers are S-invariant this point is an equilibrium point for X_S. If the fiber is connected and one dimensional then, because of S^1-symmetry, it must be a topological circle: either a periodic solution of X_S or a circle of critical points. If we have a connected part of a two dimensional critical fiber then this part is a variety whose singular set is just the critical locus. This singular set is a point or one dimensional. Therefore the former remarks apply (also cf. Smale [1970]).

From the above discussion it is clear that the periodic and stationary solutions of the system $(\mathbb{R}^4,\omega,H)$, which are also solutions of $(\mathbb{R}^4,\omega,S)$, are the singular locus Σ_H of the map $H \times S$.

In the case of two degrees of freedom suppose we have a normal form $G \times S$ for $H \times S$. By theorem 3.19. Σ_H is diffeomorphic to Σ_G, that is, the family of common periodic solutions of $(\mathbb{R}^4,\omega,H)$ and $(\mathbb{R}^4,\omega,S)$ (also the short periodic solutions of $(\mathbb{R}^4,\omega,H)$) is diffeomorphic to the family of common periodic solutions of $(\mathbb{R}^4,\omega,G)$ and $(\mathbb{R}^4,\omega,S)$. Notice that these common periodic solutions are just the relative equilibria of $(\mathbb{R}^4,\omega,H)$.

3.20. <u>DEFINITION</u>. Two S-invariant systems $(\mathbb{R}^4,\omega,H)$ and $(\mathbb{R}^4,\omega,G)$ are *equivalent* if $H \times S$ and $G \times S$ are equivalent. Two S-invariant Hamiltonian functions are *equivalent* if the corresponding systems are.

A direct consequence of theorem 3.19. is

3.21. <u>COROLLARY</u>. If $(\mathbb{R}^4,\omega,H)$ and $(\mathbb{R}^4,\omega,G)$ are two S-invariant equivalent systems then the G-level sets are invariant sets of the system $(\mathbb{R}^4,\omega,H)$.

In the case of two degrees of freedom the systems at resonance are of particular interest. If one has a system $(\mathbb{R}^4,\omega,H)$ at resonance then it is ovious that the flow of the semisimple part S of H_2 is an

S^1-action. If H is invariant under this action, that is, if H is in Hamiltonian normal form, then we may try to find a normal form for H × S. It turns out that one can find a particular simple normal form G × S (as is shown in the next section the computations can be done in such a way that the second component S is preserved, see also Duistermaat [1983[b]]). As a consequence of theorem 3.19. the qualitative behaviour of $(\mathbb{R}^n,\omega,G)$ gives much information about the qualitative behaviour of $(\mathbb{R}^4,\omega,H)$, especially about the relative equilibria.

The construction of a versal deformation (unfolding) G_ν × S allows us to study the change of the qualitative behaviour when passing through resonance. As a consequence of proposition 3.18. and theorem 3.19. we have

3.22. THEOREM. Consider two *S*-invariant systems $(\mathbb{R}^4,\omega,G_\nu)$ and $(\mathbb{R}^4,\omega,H_\mu)$ and suppose that H_{μ_0} is equivalent to G_0. Then there is a smooth map χ between the parameter spaces, $\chi(\mu) = \nu$, $\chi(\mu_0) = 0$, such that the pull-back under χ of a fibre of $G_\nu \times S$ is diffeomorphic to a fiber of $F_\mu \times S$ by means of a μ-dependent *S*-equivariant diffeomorphism (that is, the *S*-invariance is preserved). In particular energy level sets are mapped to invariant sets and relative equilibria are mapped to relative equilibria.

3.23. DEFINITION. A Hamiltonian function G_ν obtained by constructing a versal deformation $G_\nu \times S$ of a normal form G × S is called a *standard function* for the particular resonance.

3. Computation of a standard function for the nonsemisimple 1:-1 resonance

This section is completely devoted to the computation of a standard function for the nonsemisimple 1 : -1 resonance. The method we follow also works for the other resonances in two degrees of freedom. However from the computational point of view the resonance considered in this section is the most simple one.

Let $H \in C^{\infty}(\mathbb{R}^4)$ such that $H(0) = dH(0) = 0$ and consider a system $(\mathbb{R}^4, \omega, H)$ at nonsemisimple 1 : -1 resonance. If $(x,y) = (x_1,x_2,y_1,y_2)$ are the coordinates in $\mathbb{R}^4$ then the normalized quadratic part of H is

$$(3.24) \qquad H_2(x,y) = \alpha(x_1y_2-x_2y_1) + \tfrac{1}{2}(x_1{}^2+x_2{}^2)$$

with semisimple part $\alpha S(x,y) = \alpha(x_1y_2-x_2y_1)$ and nilpotent part $X(x,y) = \frac{1}{2}(x_1{}^2+x_2{}^2)$. The Hilbert generators (see theorem 3.3.) for the algebra of polynomials invariant under the action of the one parameter group S corresponding to the flow of X_S follow directly from the theory of Hamiltonian normal forms. They are the four quadratic functions $S(x,y)$, $X(x,y)$, $Y(x,y) = \frac{1}{2}(y_1{}^2+y_2{}^2)$ and $Z(x,y) = x_1y_1 + x_2y_2$. For these functions we have the following relation:

$$(3.25) \qquad 4XY \doteq Z^2 + S^2, \; X \geqslant 0, \; Y \geqslant 0.$$

By theorem 3.4. the Taylor series at zero of an S-invariant function can be written as a series in the Hilbert generators given above. For H we have the following:

$$(3.26) \qquad H(x,y) = \alpha S + X + 4a_2Y^2 + 2a_3SY + a_1S^2 + \mathcal{O}_3(X,Y,Z,S)$$

where $\mathcal{O}_3(X,Y,Z,S)$ stand for terms in X, Y, Z and S of order three or more (cf. van der Meer [1982]). Notice that H is in H_2-normal form up to order four.

Now introduce the map $G \times S$ with G given by

$$(3.27) \qquad G(x,y) = X + aY^2 \; ; \; a^2 = 1.$$

We will use the map $G \times S$ to prove that $H \times S$ is finitely determined where H is given by (3.26) with $a_2 \neq 0$. This we do by showing that each such map $H \times S$ is in the orbit of $G \times S$.

Let $F(X,Y,Z,S)$ be the space of smooth real valued functions in X, Y, Z an S. $F_2(X,Y,Z,S)$ is the space of maps from $\mathbb{R}^4$ to $\mathbb{R}^2$ with components in $F(X,Y,Z,S)$. Let $M(X,Y,Z,S)$ be the maximal ideal in $F(X,Y,Z,S)$ considering X, Y, Z, S as variables. Furthermore let $M_2(X,Y,Z,S) = M(X,Y,Z,S) \times M(X,Y,Z,S)$. (In fact $M_2(X,Y,Z,S)$ is $M(4,2)^S$ in the variables X, Y, Z and S.) $M(X,Y,Z,S)^k$ is the k-th power of the maximal ideal, that is, the functions with Taylor series starting with terms of order k in X, Y, Z and S.

A sufficient condition for $H \times S$ to be finitely determined is that $J(G \times S)^S + M_2(G,S) \supset M(X,Y,Z,S)^2 \times M(X,Y,Z,S)$. For then, by the theory of section 1, the $\mathrm{Diff}\,(\mathbb{R}^4)_0 \times \mathrm{Diff}(\mathbb{R}^2)_0$ orbit of $G \times S$ contains $H \times S$.

In order to find $J(G \times S)^S$ we first have to determine the space of S-equivariant origin preserving vector fields $X_S(\mathbb{R}^4)_0$. Because we may consider $X_S(\mathbb{R}^4)_0$ as an $F(X,Y,Z,S)$-module we need only to find a set of generators of $X_S(\mathbb{R}^4)_0$. To simplify the computations, we introduce complex conjugate variables $z_1 = x_1 + ix_2$, $z_2 = y_1 + iy_2$, $\zeta_1 = \bar{z}_1$ and $\zeta_2 = \bar{z}_2$. In these coordinates the linear vector field X_S is given by

$$\tilde{X}_S(z,\zeta) = i\begin{pmatrix} I_2 & 0 \\ 0 & I_2 \end{pmatrix}\begin{pmatrix} z_1 \\ z_2 \\ \zeta_1 \\ \zeta_2 \end{pmatrix}$$

where I_2 is the 2×2 identity matrix. A general homogeneous vector field of order n is given by

$$\tilde{V}(z,\zeta) = \begin{bmatrix} \sum\limits_{|\alpha+\beta|=n} c^{(1)}_{\alpha\beta}\, z^\alpha\, \zeta^\beta \\ \sum\limits_{|\alpha+\beta|=n} c^{(2)}_{\alpha\beta}\, z^\alpha\, \zeta^\beta \\ \sum\limits_{|\alpha+\beta|=n} c^{(3)}_{\alpha\beta}\, z^\alpha\, \zeta^\beta \\ \sum\limits_{|\alpha+\beta|=n} c^{(4)}_{\alpha\beta}\, z^\alpha\, \zeta^\beta \end{bmatrix}$$

(with the notation conventions as in subsection 3.1). The generators of the $F(X,Y,Z,S)$-module $X_S(\mathbb{R}^4)_0$ are those vector fields $\tilde{V}$ which do not contain a factor in $M(X,Y,Z,S)$ and for which $[\tilde{V}(z,\zeta), \tilde{X}_S(z,\zeta)] = D\tilde{V}(z,\zeta) \cdot \tilde{X}_S(z,\zeta) - D\tilde{X}_S(z,\zeta) \cdot \tilde{V}(z,\zeta) = 0$. $[.,.]$ is the usual Lie bracket (see chapter 1). Now $[\tilde{V},\tilde{X}_S] = 0$ if and only if

$$c^{(1)}_{\alpha\beta}[(\alpha_1+\alpha_2-\beta_1-\beta_2)-1] = 0$$
$$c^{(2)}_{\alpha\beta}[(\alpha_1+\alpha_2-\beta_1-\beta_2)-1] = 0$$
$$c^{(3)}_{\alpha\beta}[(\alpha_1+\alpha_2-\beta_1-\beta_2)+1] = 0$$
$$c^{(4)}_{\alpha\beta}[(\alpha_1+\alpha_2-\beta_1-\beta_2)+1] = 0.$$

Thus all generators of $X_S(\mathbb{R}^4)_0$ are linear. We obtain

$$z_1\frac{\partial}{\partial z_1},\quad z_2\frac{\partial}{\partial z_1},\quad z_1\frac{\partial}{\partial z_2},\quad z_2\frac{\partial}{\partial z_2}$$

and their complex conjugates

$$\zeta_1\frac{\partial}{\partial \zeta_1},\quad \zeta_2\frac{\partial}{\partial \zeta_1},\quad \zeta_1\frac{\partial}{\partial \zeta_2},\quad \zeta_2\frac{\partial}{\partial \zeta_2},$$

as generators of $X_S(\mathbb{R}^4)_0$ (writing vector fields as differential operators). We obtain $z_1\frac{\partial}{\partial z_1}$ by taking $(\alpha,\beta) = (1,0,0,0)$, $c^{(1)}_{\alpha\beta} = 1$ and $c^{(2)}_{\alpha\beta} = c^{(3)}_{\alpha\beta} = c^{(4)}_{\alpha\beta} = 0$. The other generators are obtained in an analogous way. By taking suitable linear combinations we get the following real generators:

$$V_1(x,y) = \tilde{V}_1(z,\zeta) = i(z_1\frac{\partial}{\partial z_1} - \zeta_1\frac{\partial}{\partial \zeta_1} + z_2\frac{\partial}{\partial z_2} - \zeta_2\frac{\partial}{\partial \zeta_2})$$
$$V_2(x,y) = \tilde{V}_2(z,\zeta) = (z_1\frac{\partial}{\partial z_1} + \zeta_1\frac{\partial}{\partial \zeta_1} - z_2\frac{\partial}{\partial z_2} - \zeta_2\frac{\partial}{\partial \zeta_2})$$
$$V_3(x,y) = \tilde{V}_3(z,\zeta) = -(z_1\frac{\partial}{\partial z_2} + \zeta_1\frac{\partial}{\partial \zeta_2})$$
$$V_4(x,y) = \tilde{V}_4(z,\zeta) = z_2\frac{\partial}{\partial z_1} + \zeta_2\frac{\partial}{\partial \zeta_1}$$
$$V_5(x,y) = \tilde{V}_5(z,\zeta) = \frac{1}{2i}(z_1\frac{\partial}{\partial z_1} - \zeta_1\frac{\partial}{\partial \zeta_1} - z_2\frac{\partial}{\partial z_2} + \zeta_2\frac{\partial}{\partial \zeta_2})$$
$$V_6(x,y) = \tilde{V}_6(z,\zeta) = (z_1\frac{\partial}{\partial z_1} + \zeta_1\frac{\partial}{\partial \zeta_1} + z_2\frac{\partial}{\partial z_2} + \zeta_2\frac{\partial}{\partial \zeta_2})$$
$$V_7(x,y) = \tilde{V}_7(z,\zeta) = i(z_1\frac{\partial}{\partial z_2} + \zeta_1\frac{\partial}{\partial \zeta_2})$$
$$V_8(x,y) = \tilde{V}_8(z,\zeta) = -i(z_2\frac{\partial}{\partial z_1} - \zeta_2\frac{\partial}{\partial \zeta_1})$$

Note that in complex conjugate variables the functions X, Y, Z en S become $\tilde{X}(z,\zeta) = \frac{1}{2}z_1\zeta_1$, $\tilde{Y}(z,\zeta) = \frac{1}{2}z_2\zeta_2$, $\tilde{Z}(z,\zeta) = \frac{1}{2}(\zeta_1 z_2+z_1\zeta_2)$ and $\tilde{S}(z,\zeta) = \frac{1}{2i}(\zeta_1 z_2-z_1\zeta_2)$. Introducing the Poisson bracket $\{.,.\}$ as in definition 1.9. and the adjoint map ad on $C^\infty(\mathbb{R}^4)$ we find that $V_1 = \mathrm{ad}(S)$, $V_2 = \mathrm{ad}(Z)$, $V_3 = \mathrm{ad}(X)$ and $V_4 = \mathrm{ad}(Y)$ are just the Hamiltonia vector fields corresponding to S, Z, X and Y (written as differential operators). The action of the generators V_i on the Hilbert generators X, Y, Z and S is given in table 3.1.

	X	Y	Z	S
V_1	0	0	0	0
V_2	2X	-2Y	0	0
V_3	0	-Z	-2X	0
V_4	Z	0	2Y	0
V_5	0	0	-S	Z
V_6	X	Y	Z	S
V_7	0	S	0	2X
V_8	S	0	0	2Y

Table 3.1.
The action of the generators V_i on X, Y, Z, and S.

Thus as an $F(X,Y,Z,S)$-module $J(G \times S)^S$ is generated by the following functions:

$$E_1 = V_2\begin{pmatrix}G\\S\end{pmatrix} = \begin{pmatrix}2X-4aY^2\\0\end{pmatrix}$$

$$E_2 = V_4\begin{pmatrix}G\\S\end{pmatrix} = \begin{pmatrix}Z\\0\end{pmatrix}$$

$$E_3 = V_5\begin{pmatrix}G\\S\end{pmatrix} = \begin{pmatrix}0\\Z\end{pmatrix}$$

$$E_4 = V_6\begin{pmatrix}G\\S\end{pmatrix} = \begin{pmatrix}X+2aY^2\\S\end{pmatrix}$$

$$E_5 = V_7\begin{pmatrix}G\\S\end{pmatrix} = \begin{pmatrix}2aYS\\2X\end{pmatrix}$$

$$E_6 = V_8\begin{pmatrix}G\\S\end{pmatrix} = \begin{pmatrix}S\\2Y\end{pmatrix}.$$

Here we use vector notation for $(G \times S)(X,Y,Z,S) = (G,S)$.

Let I be the $F(X,Y,Z,S)$-module generated by E_2 and E_3. Clearly I is a submodule of $F_2(X,Y,Z,S)$. Also I is a submodule of $J(G \times S)^S$. Define $\hat{J} = J(G \times S)^S/_I$. Then $\hat{J}$ is just the $F(X,Y,S)$-module generated by E_1, E_4, E_5 and E_6. Thus we have the

3.28. LEMMA. $J(G \times S)^S + M_2(G,S) \supset M(X,Y,Z,S)^2 \times M(X,Y,Z,S)$ if and only if $\hat{J} + M_2(G,S) \supset M(X,Y,S)^2 \times M(X,Y,S)$.

On $F_2(X,Y,Z,S)$ we have the relation $4XY = S^2 + Z^2$. Thus on $F_2(X,Y,Z,S)/I = F_2(X,Y,S)$ we have the relation $4XY = S^2$.

Now let $\hat{U}(X,Y,S)$ be the projection onto the first component of the subspace of $\hat{J} + M_2(G,S)$ which contains those functions into $\mathbb{R}^2$ with zero second component, The next lemma gives a precise description of $\hat{U}(X,Y,S)$ as a subspace of $F(X,Y,S)$. Here $F(X,Y,S) . \{E\}$ denotes the $F(X,Y,S)$-module generated by E.

3.29. LEMMA. $\hat{U}(X,Y,S) = F(X,Y,S) . \{2X-4aY^2\} + F(G,S) . \{3aYS\} + F(G,S) . \{X+2aY^2\} + M(G,S)$.

proof. To determine $\hat{U}(X,Y,S)$ we have to determine all functions in $\hat{J} + M_2(G,S)$ with zero second component. First we have all functions in $\hat{J}$ with zero second component. These functions are in the $F(X,Y,S)$-module generated by E_1, $SE_4 - 2YE_5$ and $SE_4 - 2XE_6$. This gives $F(X,Y,S) . \{2X-4aY^2\} \subseteq \hat{U}(X,Y,S)$. Second we have all functions in $M_2(G,S)$ with zero second component. This gives $M(G,S) \subseteq \hat{U}(X,Y,S)$. Finally we have all functions in $\hat{J}$ with second component in $M(G,S)$ because adding an element in $M_2(G,S)$ makes the second component zero. These functions form precisely the $F(G,S)$-module generated by E_1 and $E_5 + aYE_6$. Consequently $F(G,S) . \{3aYS\} + F(G,S) . \{X+2aY^2\} \subseteq \hat{U}(X,Y,S)$. (Note that the generators $3aYS$ and $X + 2aY^2$ of the $F(G,S)$-module can also be found by considering the S-equivariant vector fields mapping S to G resp. S to S). □

A direct consequence of lemma 3.29. is

3.30. <u>LEMMA</u>. If $F \in \hat{U}(X,Y,S)$ then $SF \in \hat{U}(X,Y,S)$.

The following lemma is fundamental in the proof of (2,1)-determinacy of $G \times S$ in $F_2(X,Y,Z,S)$.

3.31. <u>LEMMA</u>. $M(X,Y,S)^2 \subset \hat{U}(X,Y,S)$.

Notice that lemma 3.31 is equivalent to the fact that the first component is two-determined. To prove this it is sufficient to show that $j_2M(X,Y,S)^2 \subset j_2\hat{U}(X,Y,S)$ where j_2 is the projection onto the two-jet. This inclusion and thus lemma 3.31. is a consequence of the following lemma which is expressed in terms of formal power series. We write $\hat{U}[[X,Y,S]]$ for the formal power series version of $\hat{U}(X,Y,S)$ and $H_m[X,Y,S]$ for the homogeneous polynomials in X, Y, S. $\mathbb{R}[[X,Y,S]]$ denotes the space of formal power series in X, Y, S. Note that because of the relation $4XY = S^2$ we have $\mathbb{R}[[X,Y,S]] = \mathbb{R}[[X,S]] + \mathbb{R}[[Y,S]]$ and $H_m[X,Y,S] =$ $= H_m[X,S] + H_m[Y,S]$.

3.32. <u>LEMMA</u>. (a) $H_m[X,S] \subset j_m\hat{U}[[X,Y,S]], \forall m \geq 1$.

(b) $H_m[Y,S] \subset \hat{U}[[X,Y,S]], \forall m \geq 2$.

<u>proof of a</u>. We have $H_{m-1}[X,Y,S] . \{2X-4aY^2\} = H_{m-1}[X,S] . \{2X-4aY^2\} +$ $+ H_{m-1}[Y,S] . \{2X-4aY^2\} \subset \hat{U}[[X,Y,S]]$. Thus $j_m(H_{m-1}[X,S] . \{2X-4aY^2\})$ $= H_{m-1}[X,S] . \{X\} \subset j_m\hat{U}[[X,Y,S]]$ for $m \geq 1$. Furthermore $S^m \subset \hat{U}[[X,Y,S]]$ for $m \geq 1$. Thus part (a) follows.

<u>proof of b</u>. (Using induction) $\mathbb{R}[[Y,S]] . \{2X-4aY^2\} \subset \hat{U}[[X,Y,S]]$ and $\mathbb{R}[[G,S]] . \{X+2aY^2\} \subset \hat{U}[[X,Y,S]]$ thus $Y^2 \in \hat{U}[[X,Y,S]]$. From $\mathbb{R}[[G,S]] . \{3aYS\} \subset \hat{U}[[X,Y,S]]$ we get $YS \in \hat{U}[[X,Y,S]]$.Trivially

$S^2 \in \hat{U}[[X,Y,S]]$. Thus $H_2[Y,S] \subset \hat{U}[[X,Y,S]]$. Suppose $H_m[Y,S] \subset \hat{U}[[X,Y,S]]$. Using this hypothesis and lemma 3.30. we see that all monomials $Y^\alpha S^\beta$, $\alpha + \beta = m + 1$, $\alpha < m + 1$, lie in $\hat{U}[[X,Y,S]]$. Finally consider $Y^{m-1}(2X-4aY^2) = 2Y^{m-2}S^2 - 4aY^{m+1} \in \mathbb{R}[[Y,S]] . \{2X-4aY^2\}$. Using the induction hypothesis we obtain $Y^{m+1} \in \hat{U}[[X,Y,S]]$ and thus $H_{m+1}[Y,S] \subset \hat{U}[[X,Y,S]]$. Thus part (b) follows by induction. Notice that the condition $a \neq 0$ is essential. (This condition is a hypothesis in (3.27.).) □

Considering $E_4 - \frac{1}{2}E_1, E_5, E_6$ and using lemma 3.31. it follows straightforwardly that

3.33. <u>LEMMA</u>. $(0, M(X,Y,S)) \subset \hat{J} + M_2(G,S)$

From lemmas 3.31., 3.33 and 3.28. we have

3.34. <u>THEOREM</u>. $M(X,Y,Z,S)^2 \times M(X,Y,Z,S) \subset J(G \times S)^S + M_2(G,S)$ (provided $a \neq 0$).

3.35. <u>THEOREM</u>. Any energy momentum map for a system in nonsemisimple 1 : -1 resonance with S-invariant Hamiltonian H as given in (3.26.) is (2,1)-determined in $M_2(X,Y,Z,S)$ (that is, (4,2)-determined in $\mathbb{R}^4$ with coordinates (x,y)) provided that the coefficient of Y^2 in H is not zero.

Moreover we have shown

3.36. <u>THEOREM</u>. The complement of $J(G \times S)^S + M_2(G,S)$ in $M_2(X,Y,Z,S)$ is the element (Y,0), that is, the co-dimension of G × S is one.

<u>proof</u>. Up until now we have shown that $J(G \times S)^S + M_2(G,S)$ contains every map with first component in $M(X,S) + M(Y,S)^2 + I$ and arbitrary second component. The complement of $M(X,S) + M(Y,S)^2 + I$ in $M(X,Y,Z,S)$ is Y. From the proof of lemma 3.32. it is clear that $Y \notin \hat{U}(X,Y,Z,S)$. This proves the theorem. □

We now state a few corollaries which are in fact reformulations of the theorems 3.34. - 3.36. using the theory of section 1 and 2.

3.37. <u>COROLLARY</u>. Suppose we have a system in nonsemisimple 1 : -1 resonance with S-invariant Hamiltonian H. If the coefficients of Y^2 in H is positive then $H \times S$ has normal form $(X+Y^2,S)$. If the coefficient of Y^2 in H is negative then $H \times S$ has normal form $(X-Y^2,S)$.

3.38. <u>COROLLARY</u>. The energy-momentum map $G \times S$ with G as in (3.27.) has universal unfolding $(X+\nu Y+aY^2,S)$.

3.39. <u>COROLLARY</u>. A standard function for the nonsemisimple 1 : -1 resonance is $G_\nu = X + \nu Y + aY^2$, $a^2 = 1$.

3.40. <u>REMARK</u>. In fact one should speak of standard functions, one for $a = 1$ and one for $a = -1$. Which one to take depends on the sign of the coefficient of Y^2 of the original Hamiltonian function.

Because ν can be considered as a detuning parameter for the resonance, we can also say that G_ν is a standard function for the Hamiltonian Hopf bifurcation. Note that in this chapter we have restricted ourselves to S-invariant functions, that is, we considered integrable versions of the nonsemisimple 1 : -1 resonance and of the Hamiltonian Hopf bifurcation. In chapter 5 we shall consider the nonintegrable cases.

4. Discussion

The following is the basic idea of this chapter: to obtain standard functions for circle symmetric Hamiltonian systems, one has to consider the energy-momentum mapping instead of the Hamiltonian function alone. Looking at energy-momentum maps one sees that right-left action is the group action which one needs to obtain a maximal reduction of the

Hamiltonian. The only condition on this action is that the group should preserve the S^1-symmetry. In other words one has to consider equivariant diffeomorphisms on the source-space of the energy-momentum mapping. The normalization to standard form is now a straight forward application of the equivariant theory of stability of maps.

Although the symmetry is preserved we have to drop the requirement that the diffeomorphisms on the source space be symplectic. As a consequence, the fibration of the map is preserved but the information about the periods of the flow is lost if one translates the normalized map back to a system with standard symplectic form. (Of course this is not the case if one considers the transformed symplectic form.) However it is very hard to obtain full information about the flow. Our main aim was to preserve full information about the geometry of the periodic solutions near the equilibrium in the case of systems with two degrees of freedom. For this preserving the symplectic structures was not needed.

The theory treated in this chapter also applies to the other two degree of freedom resonances as is shown by Duistermaat [1983[b]]. There it is shown that the idea of considering energy-momentum maps leads to a reduction of co-dimension by one from his earlier computations. (Duistermaat [1982].)

One might also express the results of this chapter in the language of singularities. Then what we did was to normalize a singularity in $C^\infty(\mathbb{R}^4,\mathbb{R}^2)$ invariant under the symplectic S^1-action generated by the flow of X_S. We require that the first component of $H \times S$ has four jet $aS + bX + cY^2 + dSY + eS^2$, $b \neq 0$, $c \neq 0$ while the second component must be S. The restriction on the first component H can be expressed by saying that H is the S-invariant Hamiltonian for a system in nonsemi-simmple $1 : -1$ resonance such that the coefficient of Y^2 in H is nonzero.

Chapter IV

The Hamiltonian Hopf bifurcation

0. Introduction

In this chapter we study the fibration of the normalized map $G_\nu \times S$. In the context of Hamiltonian mechanics we may consider this map as an energy-momentum map. Then the fibration of $G_\nu \times S$ gives a fibration of the phase space of the system $(\mathbb{R}^4, \omega, G_\nu)$ into invariant sets (cf. Smale [1978]).

The investigation of the fibers of $G_\nu \times S$ may be simplified if we consider the factorization of $G_\nu \times S$ through the Hilbert map ρ for the $\mathcal{S}$-action generated by the flow of X_S (see chapter 3 section 1). If we suppose that G_ν and S are functionally independent (which is necessary for $G_\nu \times S$ to be an energy-momentum map) then we might as well take G_ν and S as Hilbert generators. Thus $G_\nu \times S$ can be written as the composition of ρ and a projection π, i.e. $G_\nu \times S = \pi \circ \rho$. Now $\rho(\mathbb{R}^4)$ is the orbit space for the S^1-action $\mathcal{S}$ generated by the flow of X_S. Therefore we know the fibration of $G_\nu \times S$ if we know the fibration of the orbit space given by π. Also ρ can be considered as a reduction map because $\rho(S^{-1}(s))$ is the reduced phase space and $\rho_* G_\nu$ is the reduced Hamiltonian. The fibers of π on the orbit space are invariant sets for the reduced Hamiltonian system.

In the case of a two degrees of freedom system the fibers of π on the orbit space are the trajectories of the reduced system. The critical π fibers, which might be points, respectively, self intersecting loops, then give rise to elliptic, respectively, hyperbolic relative equilibria for the system $(\mathbb{R}^4, \omega, G_\nu)$. To study the periodic solutions due to the $\mathcal{S}$-action one therefore has to study the singularities of the projection π.

In the case of the Hamiltonian Hopf bifurcation it turns out that the family of relative equilibria is described by a part of the well-known swallowtail surface (see figures 4.16 and 4.17). Here one has to take care that the swallowtail surface is not confused with the swallowtail singularity. The swallowtail surface obtained in this chapter is the result of a combination of a fold and a cusp singularity (see remark 4.12.).

The phenomena described in this chapter concern the qualitative behaviour of the system $(\mathbb{R}^4, \omega, G_\nu)$ and all of them have to do with the fibration of $G_\nu \times S$. Using theorems 3.19. and 3.22. and corollary 3.21. these phenomena therefore hold for all equivalent systems (see definition 3.20.).

1. Symmetry and reduction

Consider a Hamiltonian system on $\mathbb{R}^{2n}$ with Hamiltonian H and independent quadratic integral S. Let S be the linear one parameter group given by the flow of X_S. Suppose that S gives a free S^1-action on $\mathbb{R}^{2n} \setminus \{0\}$, the origin being a stationary point for X_S. Furthermore consider the Hilbert generators for the algebra of S-invariant polynomials (see chapter 3 section 1). Call these generators ρ_i, $1 \leqslant i \leqslant k$, and define the Hilbert map $\rho: (x,y) \in \mathbb{R}^{2n} \to (\rho_1, \ldots, \rho_k) \in \mathbb{R}^k$.

4.1. PROPOSITION. (Poenaru [1976]). The map ρ has the following properties

(a) ρ is proper

(b) ρ separates the orbits of S

(c) The diagram commutes, ρ' being a homeomorphism.

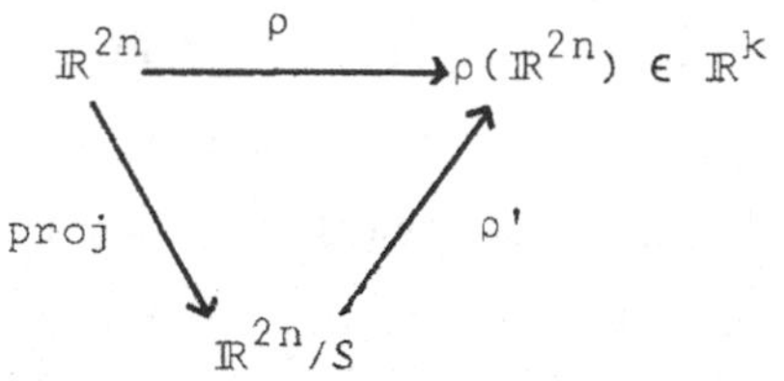

By this proposition we may consider $\rho(\mathbb{R}^{2n})$ as the orbit space for the action of S, each point of $\rho(\mathbb{R}^{2n})$ representing a different orbit. Generally $\rho(\mathbb{R}^{2n})$ is a semi-algebraic variety in $\mathbb{R}^k$ determined by the relations and inequalities satisfied by the generators. Now we may always choose S itself as one of the Hilbert generators. Then the variety $S^{-1}(s)$ is mapped by ρ onto a semi-algebraic variety $\rho(S^{-1}(s))$ in $\mathbb{R}^{k-1}$. To be more precise $\rho(S^{-1}(s))$ is the intersection of $\rho(\mathbb{R}^4)$ with the hyperplane $S = s$ in $\mathbb{R}^k$. We may consider $\rho: S^{-1}(s) \to \rho(S^{-1}(s))$, $s \neq 0$, and $\rho: S^{-1}(0)\setminus\{0\} \to \rho(S^{-1}(0))\setminus\{0\}$ as a principle S-bundle.

Let $J: \mathbb{R}^{2n} \to \mathfrak{s}^*$ be the momentum map for the action S. ($\mathfrak{s}^*$ is the dual of the Lie algebra of S). Then $J^{-1}(\mu)/S$, $\mu \in \mathfrak{s}^*$ a regular value of J, is called the reduced phase space. There exists a unique symplectic form on the reduced phase space (see theorem 1.21.). If μ is not a regular value then the theory goes through but one has to consider $J^{-1}(\mu)$ without its critical set. Following Kummer [1981] we have

4.2. <u>PROPOSITION</u>. The spaces $\rho(S^{-1}(s))$, $\rho(S^{-1}(0))\setminus\{0\}$ and the reduced phase spaces $J^{-1}(\mu)/S$ are symplectomorphic.

Thus we may take $\rho(S^{-1}(s))$, $s \neq 0$ and $\rho(S^{-1}(0)\setminus\{0\}$ as a model for the reduced phase spaces. Let $M_s = \rho(S^{-1}(s))$ then the reduced phase spaces are M_s and $M_0\setminus\{0\}$. Now ρ_*H is the reduced Hamiltonian defining a reduced system on M_s, respectively, M_0, the origin on M_0 being a stationary point for the reduced system.

The above reduction can be performed for all two degree of freedom resonances if one supposes the Hamiltonian to be invariant under the action of the flow of the semisimple part S of the linearized system (cf. Cushman and Rod [1982], Cushman [1983,1984], Churchill et al [1983]).

Now consider the case of the $1 : \pm 1$ resonances. Let G be the group of linear symplectic transformations that leave S invariant and let $\mathfrak{g}$ be the corresponding Lie algebra. Only in these cases can the Hilbert generators be chosen to be homogeneous quadratic polynomials. The set of generators under the Poisson bracket forms a Lie algebra which can be identified with the dual $\mathfrak{g}^*$ of $\mathfrak{g}$. In these cases one may establish that the reduced phase spaces are coadjoint orbits of G on $\mathfrak{g}^*$ (Cushman and Rod [1982], Cushman [1982[a,b]]).

We will treat the Hamiltonian Hopf bifurcation in detail. In chapter 3 section 3 we found that the Hilbert generators are the four homogeneous quadratic polynomials $X(x,y) = \frac{1}{2}(x_1^2+x_2^2)$, $Y(x,y) = \frac{1}{2}(y_1^2+y_2^2)$, $Z(x,y) = x_1y_1 + x_2y_2$ and $S(x,y) = x_1y_2 - x_2y_1$. ($(x,y) = (x_1,x_2,y_1,y_2)$ being the coordinates on $\mathbb{R}^4$). Of course we may also take $Y(x,y)$, $Z(x,y)$, $S(x,y)$ and $G_\nu(x,y) = X(x,y) + \nu Y(x,y) + aY^2(x,y)$ as Hilbert generators. In fact this change of generators is nothing more than a diffeomorphism of the target space of ρ. The homogeneous generators satisfy the relation $4XY = Z^2 + S^2$, $X \geqslant 0$, $Y \geqslant 0$ which is equivalent to

$$Z^2 + S^2 + 4\nu Y^2 + 4aY^3 - 4YG_\nu = 0,\ Y \geqslant 0,\ \nu Y + aY^2 - G_\nu \leqslant 0. \tag{4.3}$$

The image $M = \rho(\mathbb{R}^4)$ of the new Hilbert map $\rho\colon \mathbb{R}^4 \to \mathbb{R}^4$ defined by $(x_1,x_2,y_1,y_2) \mapsto (G_\nu,Y,Z,S)$ is given by equation (4.3), that is,

$$M = \{(G_\nu,Y,Z,S) \in \mathbb{R}^4 \mid Z^2+S^2+4\nu Y^2+4aY^3-4YG_\nu=0,\ Y \geqslant 0,\ \nu Y+aY^2-G_\nu \leqslant 0\}.$$

M is a deformed half cone with vertex at the origin. Let M_s be an S-slice of M, that is,

$$M_s = \{(G_\nu,Y,Z) \in \mathbb{R}^3 \mid Z^2+4\nu Y^2+4aY^2-4YG_\nu = -s^2,\ Y \geqslant 0,\ \nu Y+aY^2-G_\nu \leqslant 0\}$$

As before the reduced phase spaces are M_s and $M_0\setminus\{0\}$. Notice that $M_s = M_{-s}$. Furthermore on each M_s we have an additional symmetry induced by the antisymplectic reflection R on $\mathbb{R}^4$ defined by

$$R: (x_1,x_2,y_1,y_2) \mapsto (x_1,-x_2,-y_1,y_2)$$

The energy-momentum map $G_\nu \times S$ is invariant under R. ρ_*R gives a reflection on the target space M of ρ defined by

$$\rho_*R: (G_\nu,Y,Z,S) \mapsto (G_\nu,Y,-Z,S)$$

Clearly M_s is invariant under ρ_*R.

We state this symmetry because it will be helpful later in visualizing the fibration of $G_\nu \times S$. They will be explicitly used in establishing the reduced potentials and the reduced vector fields in the next section.

2. The fibres of $G_\nu \times S$.

In this section we will determine the fibres of $G_\nu \times S$. Hereto it is sufficient to determine the reduced fibres, that is, the fibres of

$$\pi: (G_\nu,Y,Z,S) \mapsto (G_\nu,S)$$

These fibres are just the trajectories $\gamma_{g,s}$ of the reduced vector field with energy function G_ν. More precisely the trajectories $\gamma_{g,s}$ are the $G_\nu = g$ level sets on M_s. Because M_s is topologically contractible it follows that the fibres of $G_\nu \times S$ in $\mathbb{R}^4$ are $\gamma_{g,s} \times S^1$. The topological nature of $\gamma_{g,s}$ can be found by applying Morse theory to the function G_ν on M_s.

Recall that M_s is symmetric with respect to ρ_*R. Thus we only need to look at the Z = 0 slice of M_s. Now the movement of the reduced vector field takes place on G_ν-hyperplanes in (G_ν,Y,Z)-space. Considering $M_s|_{Z=0}$ as a graph of G_ν therefore gives us the potential of the reduced system. If $s \neq 0$ we obtain from (4.3) the potential function

$V_s(Y) = \nu Y + aY^2 + s^2/4Y$, $y \geqslant 0$; while for $s = 0$ we obtain $V_0(Y) = \nu Y + aY^2$ together with the positive half of the $Y = 0$ axis. In figures (4.1)-(4.14) we give characteristic examples of the graph of potential V_s (figures (4.1^a)-(4.14^a)) together with the corresponding trajectories of the reduced vector field projected onto (Y,Z)-plane (figures (4.1^b)-(4.14^b)). The topological type of the $\gamma_{g,s}$ is now immediately clear. We have $(G_\nu \times S)^{-1}(g,s) = \gamma_{g,s} \times S^1$ outside the origin. Because the action S is free outside the origin and has a stationary point at the origin we find that every fibre which contains the origin has a singularity at the origin.

On the next pages:

fig. (4.1^a) - (4.14^a): The potentials $V_s(Y)$

fig. (4.1^b) - (4.14^b): The trajectories of the reduced vector fields, that is, the G_ν-level sets on M_s, corresponding to the potentials of the a-figures. The trajectories are projected perpendicularly on the (Y,Z)-plane. (As a consequence all $G_\nu > 0$ levels pass through the origin if $s = 0$. Note that only the $G_\nu = 0$ level contains the origin in (G_ν,Y,Z)-space.)

The parameter range for which the figures are characteristic is indicated.

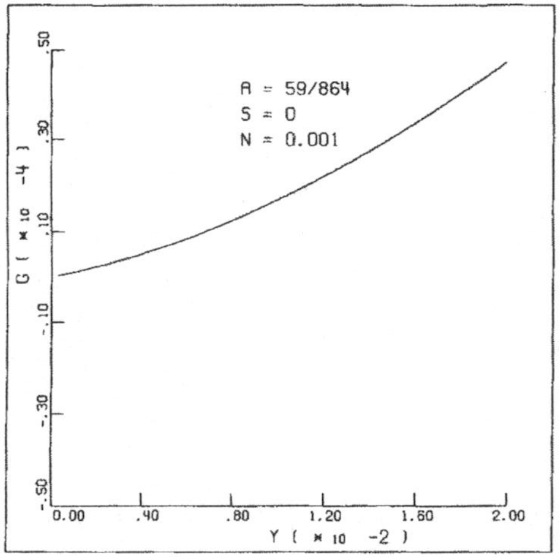

fig. (4.1[a])

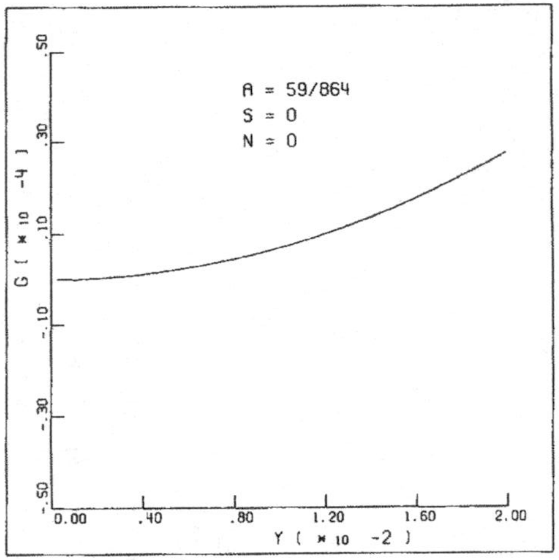

fig. (4.2[a])

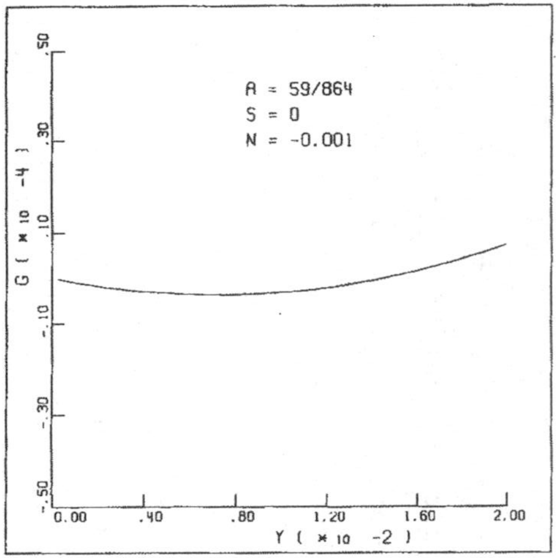

fig. (4.3[a])

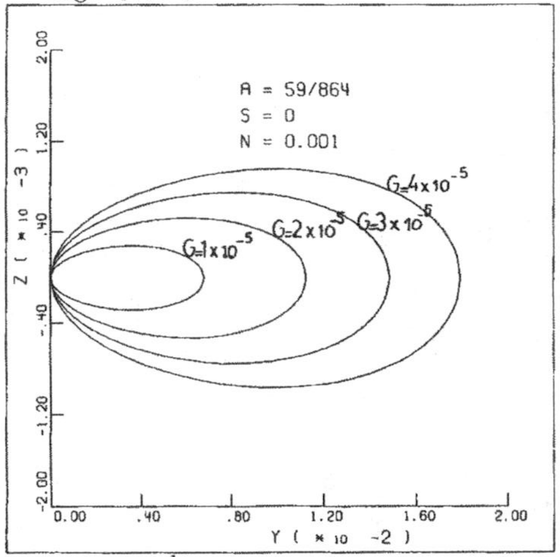

fig. (4.1[b]) $a>0$, $S=0$, $\nu>0$

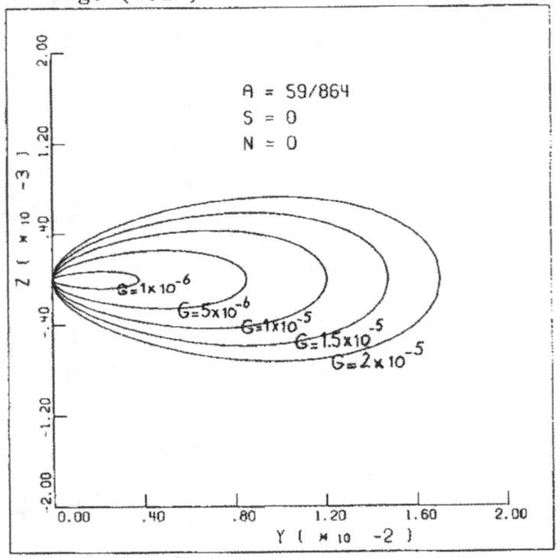

fig. (4.2[b]) $a>0$, $S=0$, $\nu=0$

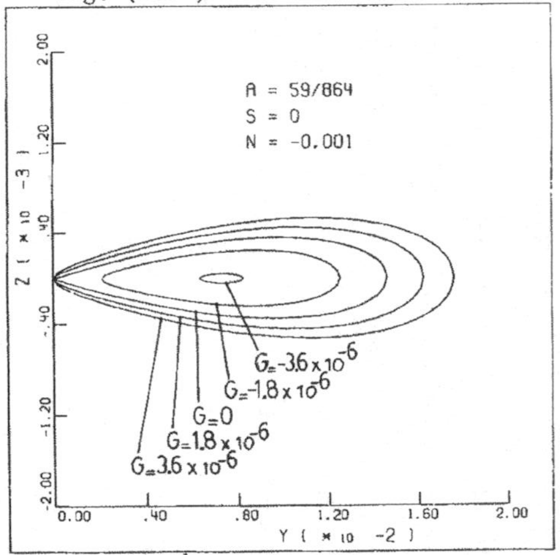

fig. (4.3[b]) $a>0$, $S=0$, $\nu<0$

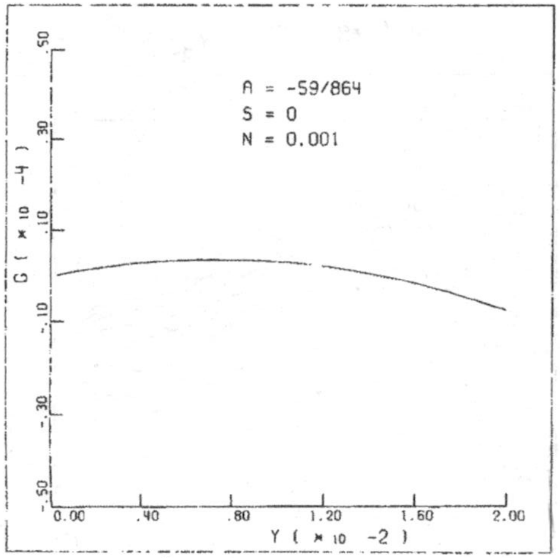

fig. (4.4^a)

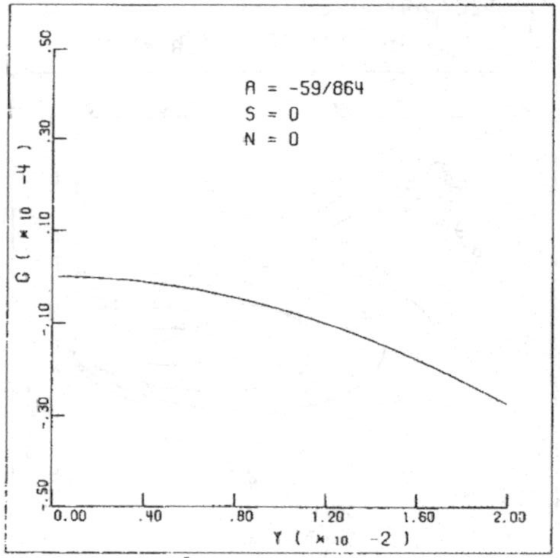

fig. (4.5^a)

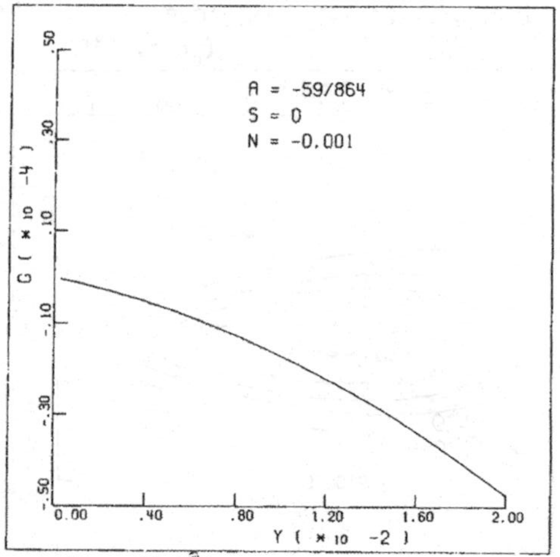

fig. (4.6^a)

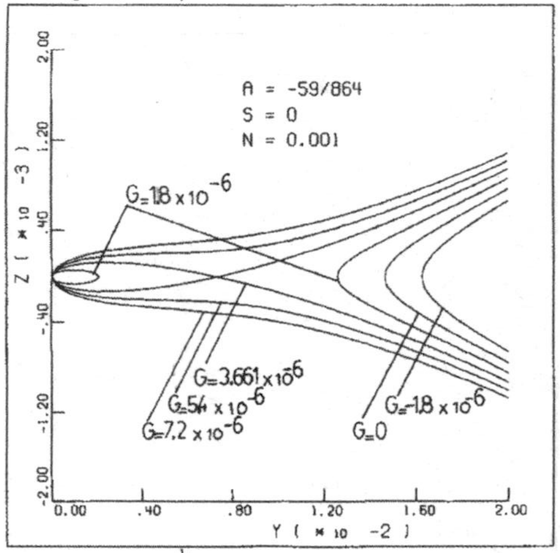

fig. (4.4^b) $a<0,\ S=0,\ \nu>0$

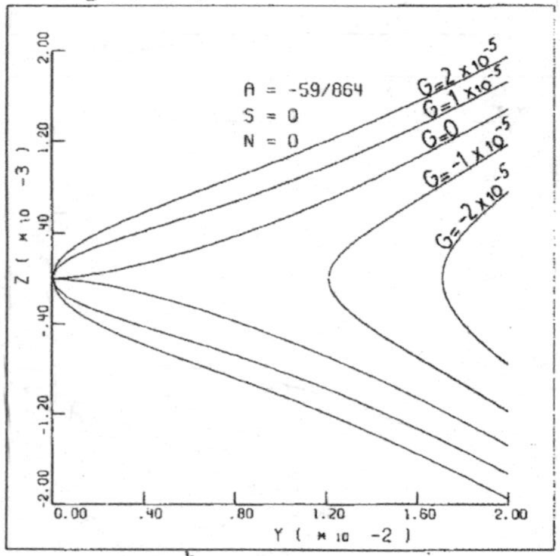

fig. (4.5^b) $a<0,\ S=0,\ \nu=0$

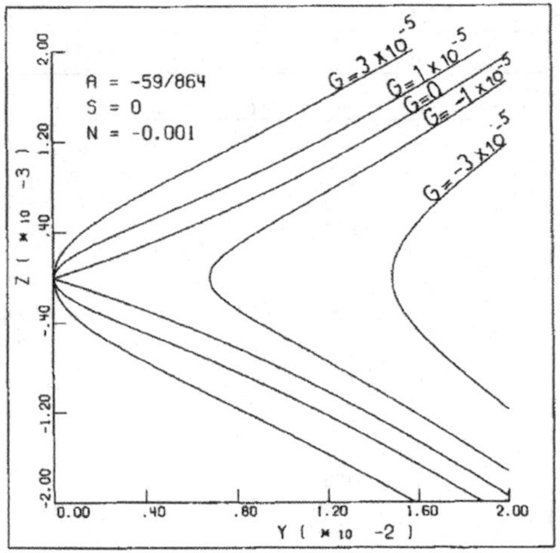

fig. (4.6^b) $a<0,\ S=0,\ \nu<0$

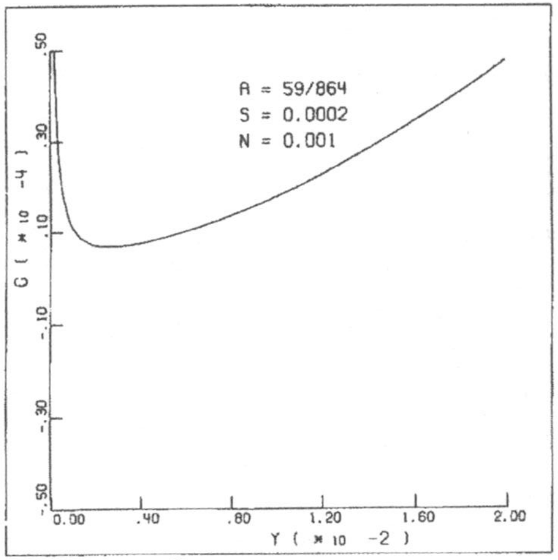

fig. (4.7^a)

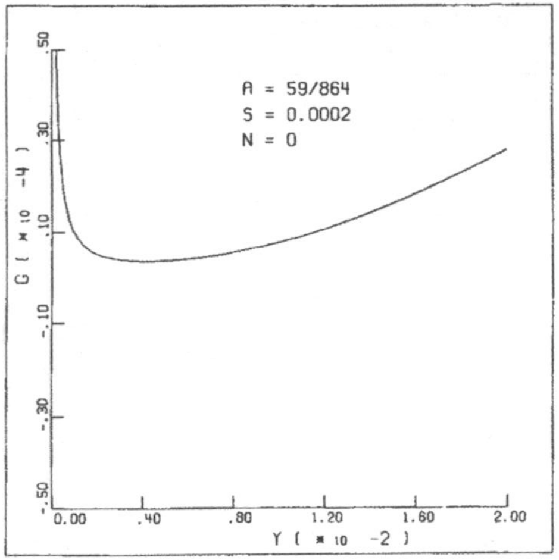

fig. (4.8^a)

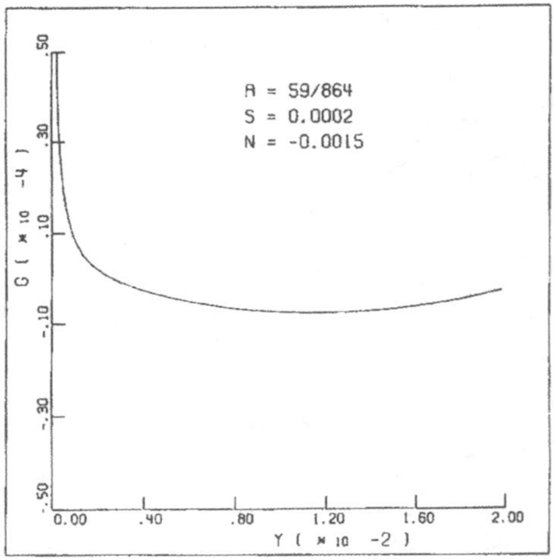

fig. (4.9^a)

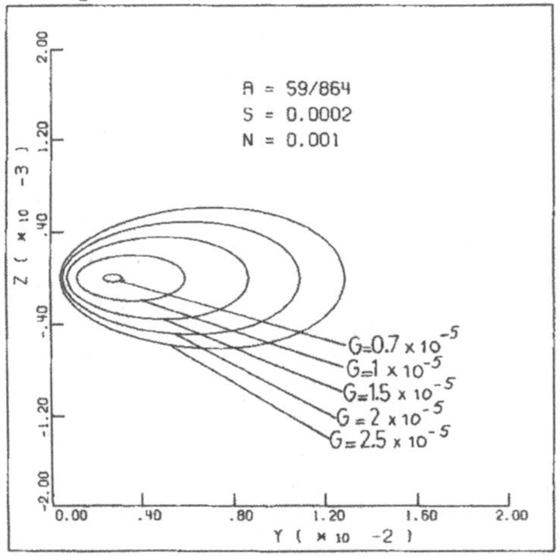

fig. (4.7^b) $a>0$, $S\neq 0$, $\nu>0$

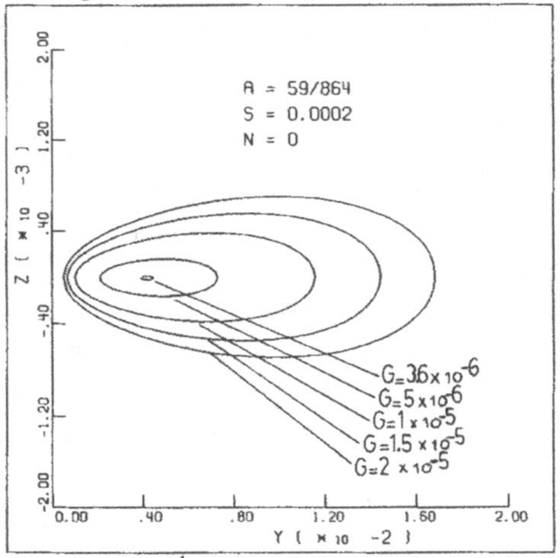

fig. (4.8^b) $a>0$, $S\neq 0$, $\nu=0$
$a>0$, $|S|>(-16\nu^3/27a^2)^{\frac{1}{2}}$, $\nu<0$

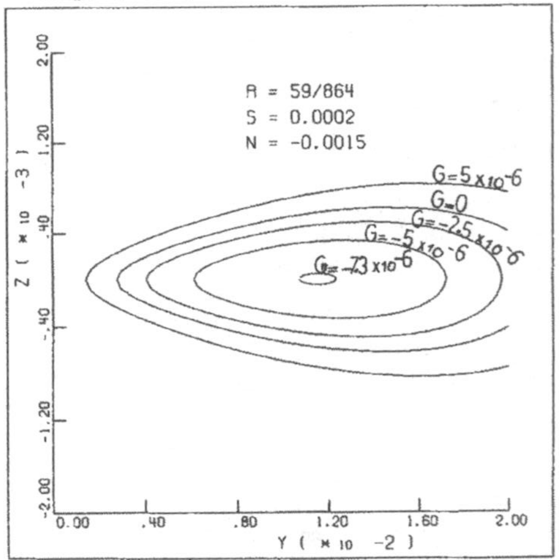

fig. (4.9^b)
$a>0$, $|S|<(-16\nu^3/27a^2)^{\frac{1}{2}}$, $\nu<0$

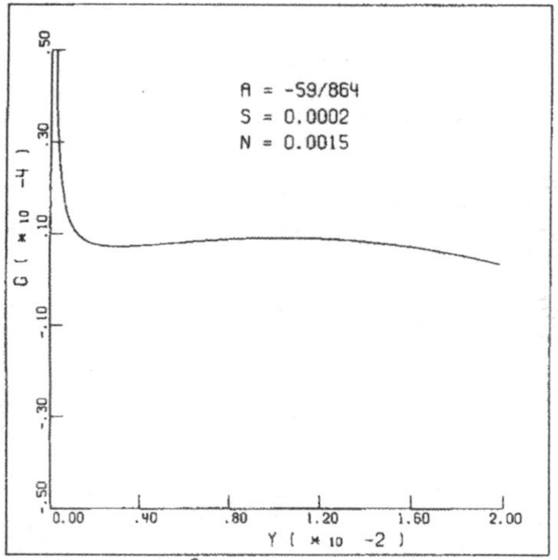

fig. (4.10[a])

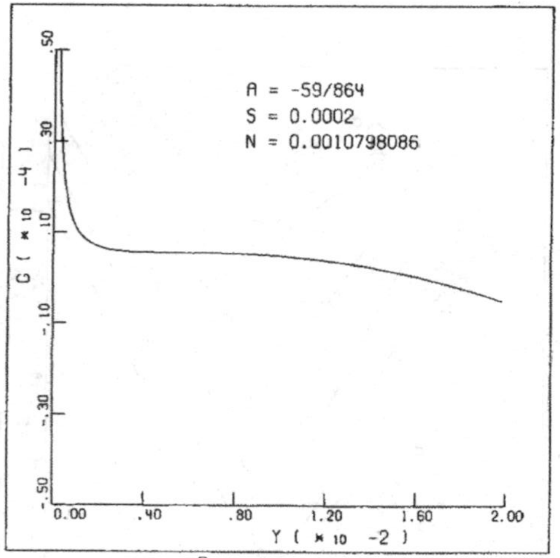

fig. (4.11[a])

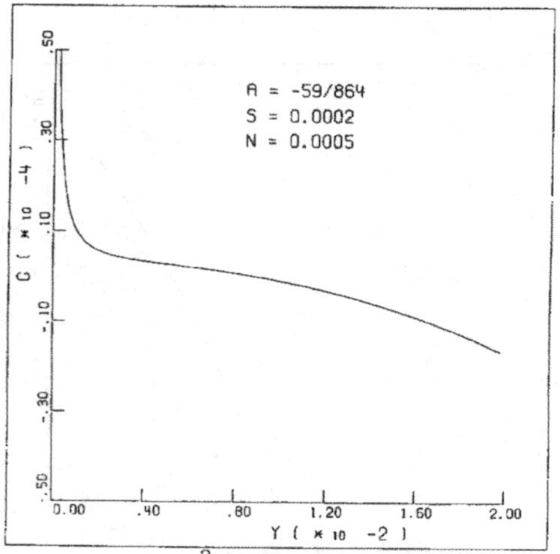

fig. (4.12[a])

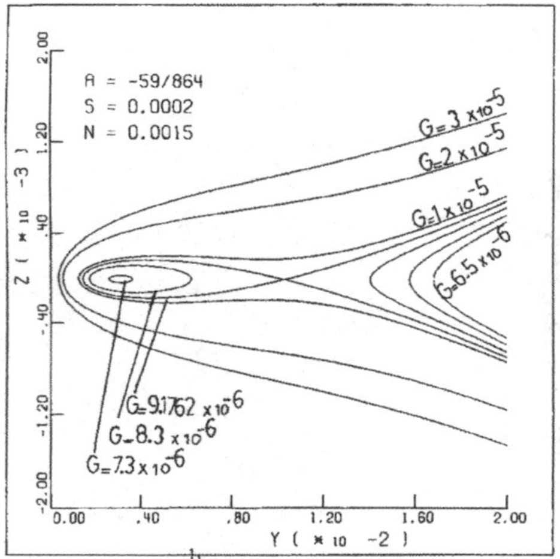

fig. (4.10[b])
$a<0, \ |S|<(-16\nu^3/27a^2)^{\frac{1}{2}}, \ \nu>0$

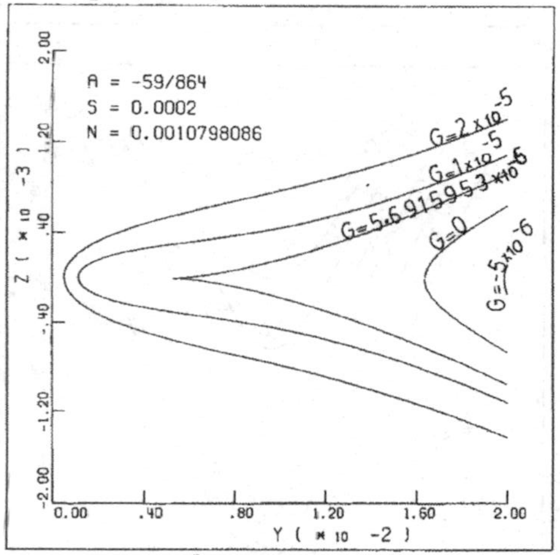

fig. (4.11[b])
$a<0, \ |S|=(-16\nu^3/27a^2)^{\frac{1}{2}}, \ \nu>0$

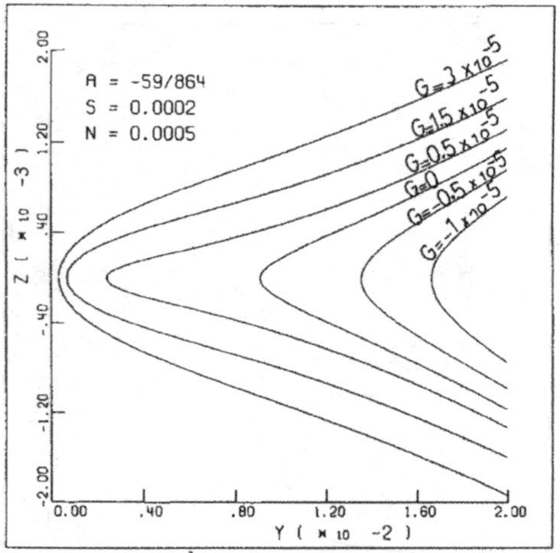

fig. (4.12[b])
$a<0, \ |S|>(-16\nu^3/27a^2)^{\frac{1}{2}}, \ \nu>0$

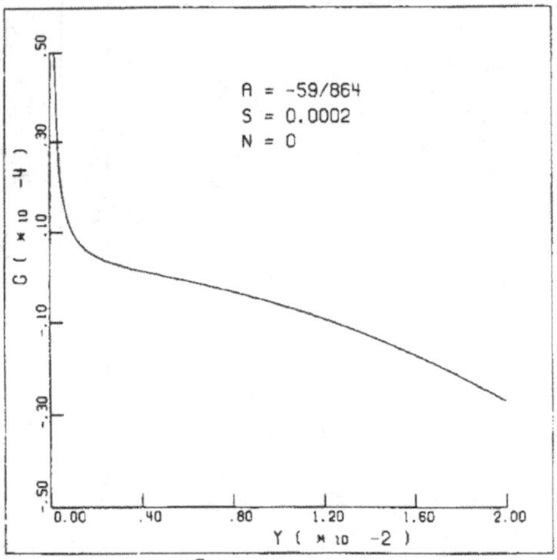

fig. (4.13[a])

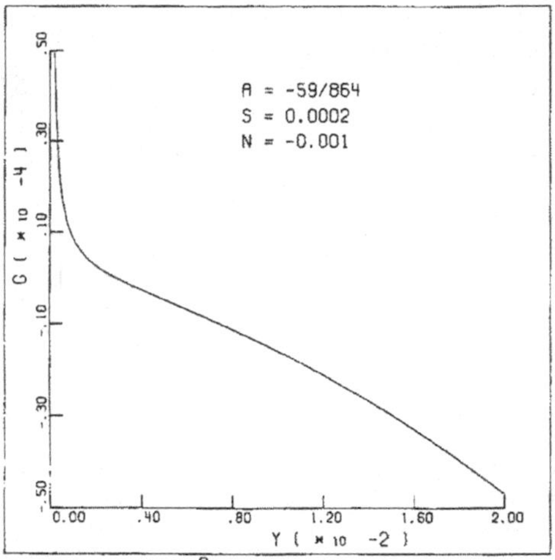

fig. (4.14[a])

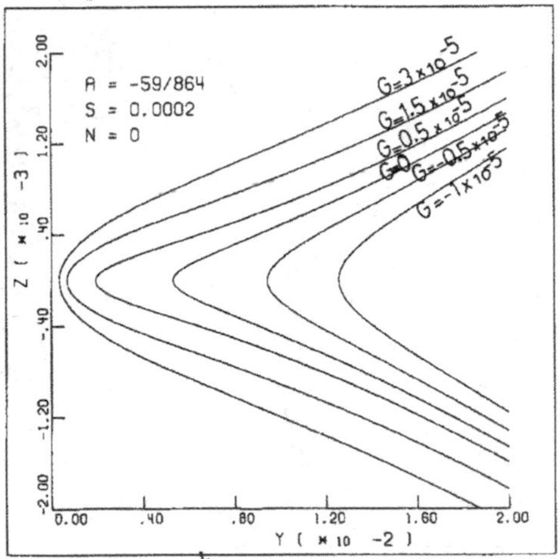

fig. (4.13[b]) $a<0,\ S\neq 0,\ \nu=0$

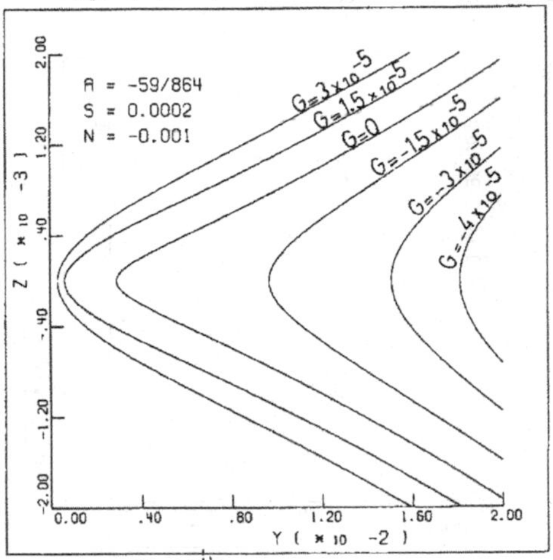

fig. (4.14[b]) $a<0,\ S\neq 0,\ \nu<0$

3. Relative equilibria

The relative equilibria of the system $(\mathbb{R}^4,\omega,G_\nu)$ are the critical points of the corresponding reduced system on M_S. By proposition 1.25. these relative equilibria form the singular locus of the energy-momentum map $G_\nu \times S$. On the image M of ρ this singular locus is characterized by the extrema of the $V_s(Y)$, that is, by the extrema of M restricted to sections S = constant and Z = 0 and considered as a graph of G_ν. Let

$$F(Y,Z,g,s,\nu) = 4aY^3 + 4\nu Y^2 - 4gY + Z^2 + s^2$$

Then M is defined by F = 0 together with $Y \geqslant 0$, $\nu Y + aY^2 - g \leqslant 0$. The relative equilibria on M are therefore given by

$$(4.4) \qquad \begin{aligned} &F(Y,0;g,s,\nu) = 0 \\ &\frac{dF}{dY}(Y,0;g,s,\nu) = 0 \ ; \ Y \geqslant 0,\ \nu Y + aY^2 - g \leqslant 0 \end{aligned}$$

which is equivalent to

$$(4.5) \qquad 4aY^3 + 4\nu Y^2 - 4gY + s^2 = 0,$$

$$(4.6) \qquad 12aY^2 + 8\nu Y - 4g = 0,$$

$$(4.7) \qquad Y \geqslant 0,\ \nu Y + aY^2 - g \leqslant 0$$

Let E be the set of points in parameter space (g,ν,s) for which (4.5), (4.6) have common solutions. Then E is the set of (g,ν,s) for which (4.5) has zeroes with multiplicity greater than one. Thus E is the discriminant locus $\Delta_5 = 0$, where Δ_5 is the discriminant of (4.5)

4.8. <u>LEMMA</u>. Consider the equations

$$(4.9) \qquad Y^4 - \frac{\nu}{2a}Y^2 + \frac{s}{\sqrt{a}}Y + \frac{g}{4a} + \frac{\nu^2}{16a^2} = 0 \ ; \ a > 0$$

$$(4.10) \qquad Y^4 + \frac{\nu}{2a}Y^2 + \frac{s}{\sqrt{-a}}Y + \frac{g}{4a} + \frac{\nu^2}{16a^2} = 0 \ ; \ a < 0.$$

Then $\Delta_5 = \Delta_9$ if $a > 0$ and $\Delta_5 = \Delta_{10}$ if $a < 0$.

The proof of this lemma is obtained by just writing down the equations for the discriminant. □

The discriminant locus of (4.9) ((4.10)) is just the well known swallowtail surface (see Poston and Stewart [1978]) given in fig. (4.15) for $a > 0$. If $a < 0$ the picture should be reflected in the (s,ν)-plane.

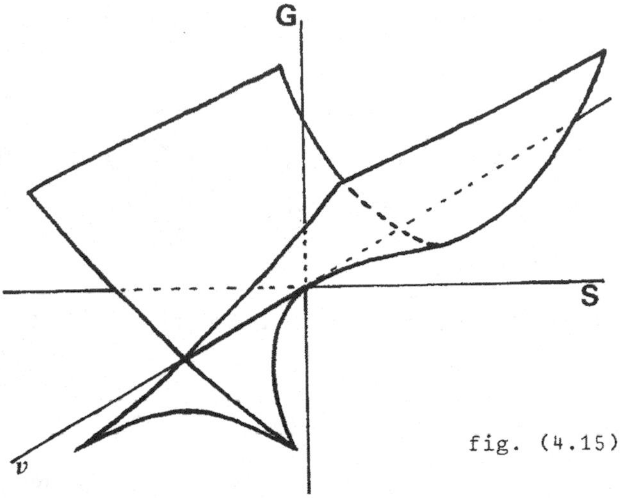

fig. (4.15). The swallowtail surface.

Because of the inequalities (4.7) the relative equilibria form a part $\mathcal{E}$ of the swallowtail E. From (4.7) we obtain the following restrictions: (1) $a > 0$, $\nu > 0$ then $g \geqslant 0$; (2) $a > 0$, $\nu < 0$ then $g \geqslant -\frac{\nu^2}{4a}$; (3) $a < 0$, $\nu > 0$ then $g \geqslant 0$ and (4) $a < 0$, $\nu < 0$ then $g \geqslant -\frac{\nu^2}{4a}$. Therefore we have

4.11. THEOREM. $\mathcal{E}$ is the part of the swallowtail surface indicated in fig. (4.16) for $a > 0$ and in fig. (4.17) for $a < 0$.

Taking s and ν constant for each g the number of positive roots of (4.5) can be read off from the graph of $V_s(Y)$. This root number is directly related to the nature of the corresponding fibre $(G_\nu \times S)^{-1}(g,s)$ as shown in table 4.1. In figures (4.18) and (4.19) the root number is indicated in sections ν = constant of E. Also the stability type of the relative equilibria is indicated with e standing for elliptic (stable), h standing for hyperbolic (unstable) and t standing for transitional. These are obtained from the potential V_s; a maximum corresponds to an unstable and a minimum to a stable relative equilibrium. Table 4.1 also describes the fibres corresponding to the critical values of $G_\nu \times S$. These fibres are or contain the relative equilibria.

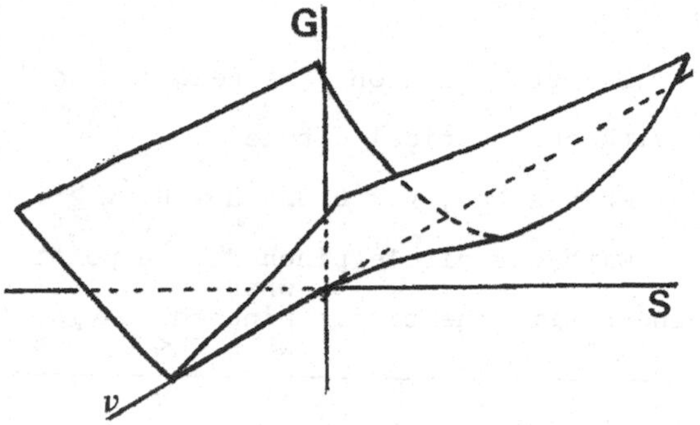

fig. (4.16). E for $a > 0$.

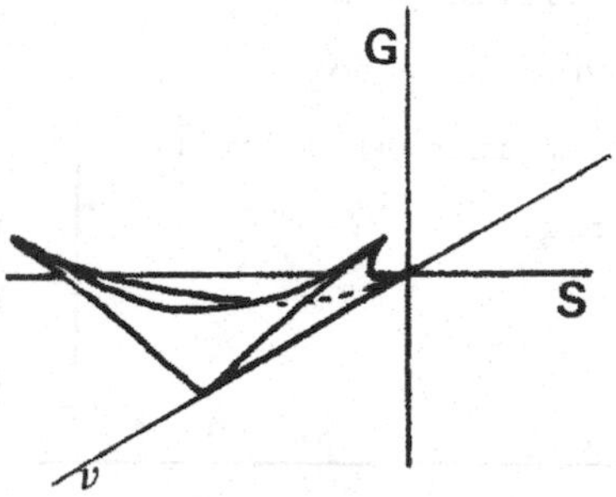

fig. (4.17). E for $a < 0$.

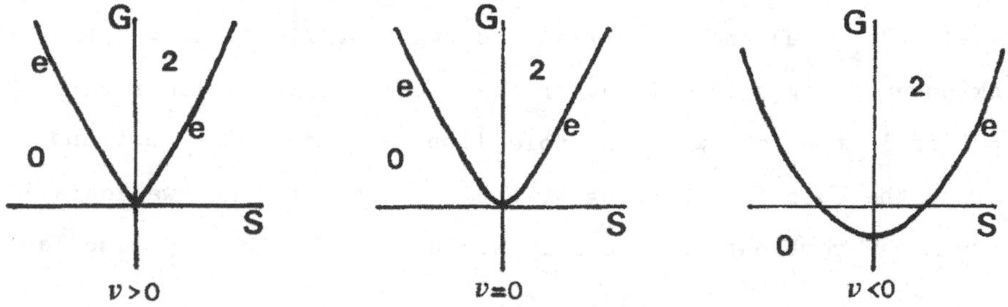

fig. (4.18). ν = constant slices with rootnumbers for $a > 0$.

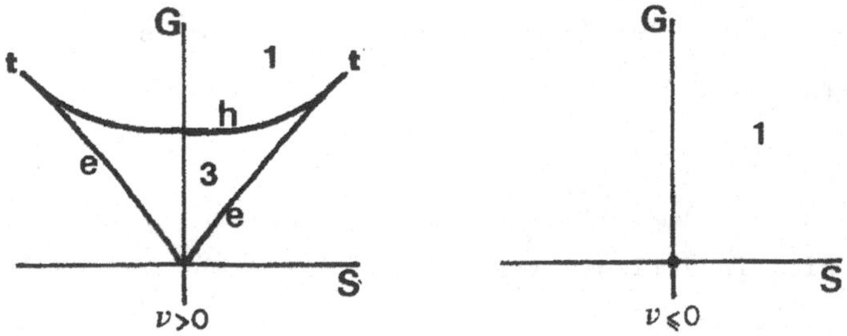

fig. (4.19). ν = constant slices with root numbers for $a < 0$.

table 4.1. The fibers of $G_\nu \times S$.

(g,s)	$(G_\nu \times S)^{-1}(g,s)$
root number 0	$\emptyset$
root number 1	$S^1 \times \mathbb{R}$
root number 2	T^2
root number 3	$T^2 \amalg S^1 \times \mathbb{R}$ ($\amalg$ means disjoint union)
elliptic point (e)	S^1
hyperbolic point (h)	$T^2 \vee S^1 \times \mathbb{R}$ attached along the relative equilibrium
transitional point (t)	$S^1 \times \mathbb{R}$ with one critical circle
origin	(1) point when $a > 0$, $\nu \geqslant 0$ and $a < 0$, $\nu > 0$ (2) torus with one circle pinched to a point when $a > 0$, $\nu \leqslant 0$ (3) cylinder with one circle pinched to a point when $a < 0$, $\nu \leqslant 0$

4.12. REMARK. The above results can also be obtained by transforming (4.5) into the standard form $Z^3 + pZ + q = 0$ by putting $Y = Z - \nu/a$. Then $p = -(\nu^2/3a^2+g/a)$ and $q = g\nu/3a^2 + s^2/4a + 2\nu^3/27a^3$. $\Delta_5 = 0$ is the pull-back under Γ: $(g,s,\nu) \mapsto (p,q)$ of the usual cusp. On planes ν = constant Γ is just a fold with the fold line tangent to the cusp and moving along the cusp for changing ν. Thus one obtains the swallowtail surface from the combination of a fold singularity and a cusp singularity.

4.13. REMARK. Theorem 4.11. describes the set of singular values of the map $G_\nu \times S$. Because $G_\nu \times S$ is the universal unfolding of a normal form, that is, a standard function, this result is generic in the class of $\mathbf{S}$-invariant energy-momentum maps.

4. The $G_\nu(x,y)$ level sets

In this section we will investigate the nature of the energy levels of the system $(\mathbb{R}^4,\omega,G_\nu)$. By corollary 3.21 this gives us information about all equivalent systems.

We will first consider the reduced level sets $M_g = \rho(G_\nu^{-1}(g))$. By equation (4.3)

$$M_g = \{(Y,Z,S) \in \mathbb{R}^3 \mid Z^2+S^2+4\nu Y^2+4aY^3-4gY = 0,\ Y \geq 0,\ \nu Y+aY^2-g \leq 0\}$$

Thus the surface M_g in (Y,Z,S)-space is obtained by rotating around the Y-axis the part of the graph of $4g - 4\nu Y^2 - 4aY^2$ defined by $Y \geq 0$, $\nu Y + aY^2 - g \leq 0$. If one adds an S-axis to figures (4.1^b) - (4.6^b) one can visualize the surfaces M_g by rotating the G-level lines around the Y-axis. The results for the different parameter values are listed in table 4.2.

table 4.2. The reduced G_ν level sets

	conditions on the parameters	M_g
I.a	$a > 0,\ \nu \geq 0,\ g < 0$	$\emptyset$
I.b	$,\ g = 0$	point
I.c	$,\ g > 0$	S^2
II.a	$a > 0,\ \nu < 0,\ g < -\nu^2/4a$	$\emptyset$
II.b	$,\ g = -\nu^2/4a$	point
II.c	$,\ -\nu^2/4a < g < 0$	S^2
II.d	$,\ g = 0$	sphere with singular point
II.e	$,\ g > 0$	S^2
III.a	$a < 0,\ \nu > 0,\ g < 0$	$\mathbb{R}^2$
III.b	$,\ g = 0$	point $\amalg$ $\mathbb{R}^2$
III.c	$,\ 0 < g < -\nu^2/4a$	$S^2 \amalg \mathbb{R}^2$
III.d	$,\ g = -\nu^2/4a$	$S^2 \vee \mathbb{R}^2$
III.e	$,\ g > -\nu^2/4a$	$\mathbb{R}^2$
IV.a	$a < 0,\ \nu \leq 0,\ g = 0$	$\mathbb{R}^2$ with singular point
IV.b	$,\ g \neq 0$	$\mathbb{R}^2$

We will now state some facts about the G(x,y) level sets in $\mathbb{R}^4$. If we consider ρ: $\mathbb{R}^4\setminus\{0\} \to \rho(\mathbb{R}^4)\setminus\{0\}$ as a principal S-bundle then the $G^{-1}(g)$ are S^1 bundles over S^2 in the cases I.c, II.c and II.e. In case III.c $G_\nu^{-1}(g)$ contains an S^1 bundle over S^2. In case II.c the S^2 are contractible in $\rho(\mathbb{R}^4)\setminus\{0\}$. Thus $G_\nu^{-1}(g)$ is the trivial bundle $S^1 \times S^2$. The origin is a nondegenerate critical point for $G_\nu(x,y)$ if $\nu > 0$. By Morse theory it follows that for g small, $g \neq 0$, $G_\nu^{-1}(g)$ has a compact component diffeomorphic to S^3. Now the critical G(x,y) levels are those for which also $\rho(G^{-1}(g))$ is critical. Thus $g = 0$ and $g = -\nu^2/4a$ are critical values of G_ν. Therefore changing ν gives rise to an isotopy of G_ν level sets (keeping g sufficiently small in case III.c). Thus the S^1-bundle $G_\nu^{-1}(g)$ must be S^3 in the cases I.c, II.e and III.c. We obtain the $G_\nu(x,y)$ level sets listed in table 4.3 using table 4.2 and the above remarks.

table 4.3. The $G_\nu(x,y)$ level sets

	$G_\nu^{-1}(g)$
I.a	$\emptyset$
I.b	point
I.c	S^3
II.a	$\emptyset$
II.b	circle of critical points
II.c	$S^2 \times S^1$
II.d	S^1 bundle over S^2 with one fibre pinched to a point
II.e	S^3
III.a	$\mathbb{R}^2 \times S^1$
III.b	(point) $\amalg$ $\mathbb{R}^2 \times S^1$
III.c	$S^3 \amalg \mathbb{R}^2 \times S^1$
III.d	$S^3 \vee \mathbb{R}^2 \times S^1$ attached along S^1
III.e	$\mathbb{R}^2 \times S^1$
IV.a	S^1 bundle over $\mathbb{R}^2$ with one fibre pinched to a point
IV.b	$\mathbb{R}^2 \times S^1$

The information about the topology of the energy levels allows us to draw some straight forward conclusions. Because in case II.c, d, e the topology of $G_\nu^{-1}(g)$ changes if g passes through zero we have

4.14. <u>COROLLARY</u>. In the case $(a > 0,\ \nu < 0)$ there is nontrivial monodromy, that is, the system admits no global action-angle coordinates in a full neighbourhood of the origin.

For more details on the subject of monodromy see Duistermaat [1980] and Cushman [1983].

4.15. <u>COROLLARY</u>. When $(a > 0,\ \nu \geq 0)$ or $(a < 0,\ \nu \geq 0)$ the origin is a Liapunov-stable stationary point for the system $(\mathbb{R}^4, \omega, G_\nu)$. In the case $(a > 0,\ \nu < 0)$ the origin is a Liapunov-unstable but solutions starting near zero are bounded. In the case $(a < 0,\ \nu \leq 0)$ the origin is Liapunov-unstable and solutions run off to infinity.

<u>proof</u>. The first statement follows from the fact that for g small $G_\nu^{-1}(g)$ has a component S^3 whose interior contains the origin, moreover $G_\nu^{-1}(g)$ pulls into the origin for $g \to 0$. The second statement follows from the fact that in this case $G_\nu^{-1}(g)$ is an S^3 which has the origin inside for g large enough. The third statement follows using Liapunov-function $Z(x,y)$. □

We note that the above corollaries also hold for equivalent systems.

Chapter V

Nonintegrable systems at resonance

0. Introduction

It is well known that some of the qualitative properties of an integrable system are preserved when this system is perturbed to a nonintegrable system. Conversely when considering nonintegrable systems one might construct a nearby integrable system such that the integrable system has some of the qualitative properties of the nonintegrable one. In this chapter we use the latter method to study families of periodic orbits near an equilibrium of a nonintegrable system passing through nonsemisimple 1 : -1 resonance.

Following Duistermaat [1983[b]] we will use a method which is based on ideas of Moser [1976] and Weinstein [1973,1978]. This method, called the Moser-Weinstein reduction, reduces the search for periodic solutions of a family of nonintegrable systems to the search for periodic solutions of a nearby family of integrable systems. In essence Moser-Weinstein reduction comes down to proving the existence of a function E invariant under the S^1-action of the semisimple part of the linearized system such that the periodic solutions of the Hamiltonian system with Hamiltonian E correspond to the periodic solutions of the original nonintegrable system by means of a C^k-diffeomorphism.

The last section of this chapter discusses some open problems concerning the nonintegrable phenomena of invariant tori and homoclinic orbits. These problems arise if one perturbs the system considered in chapter 4.

The main theorem concerning periodic solutions is stated in section 3. This theorem is a generalization of the results of chapter 3 and 4.

1. Normal form and deformation

Consider a system $(\mathbb{R}^4,\omega,F_\mu)$ where F_μ is a deformation (see def. 3.12) of a Hamiltonian function of a system in nonsemisimple 1 : -1 resonance. In other words, the quadratic term F_2^0 of F_0 has the normal form

$$F_2^0(x,y) = \alpha(x_1y_2-x_2y_1) \pm \tfrac{1}{2}(x_1^2+x_2^2).$$

As in the foregoing chapters we shall take the + sign.

We can consider F_μ as a perturbation of a S-invariant function if we transform F_μ to an S-normal form (up to some arbitrary order). If F_2^μ does not depend on μ then $F_2^\mu = F_2^0$ and we may apply the normal form theorem of chapter 2 in a straight forward way. If F_2^μ does depend on μ, then it is possible that μ detunes the resonance in such a way that by changing μ other resonances occur. This has consequences for the normal form. We shall show that one can still transform F_μ to an S-normal form up to arbitrary order provided the parameter μ is restricted to some neighborhood of zero. This neighborhood shrinks with growing order of the normal form.

By a parameter dependent linear symplectic transformation depending smoothly on the parameter, F_2^μ can be transformed to

$$H_2^\nu(x,y) = F_2^0 + \nu_1\, S(x,y) + \nu_2\, Y(x,y)$$

which is a versal deformation of F_2^0 (see van der Meer [1982]). The eigenvalues of the linear vector field $X_{H_2^\nu}$ are

$$(1+\nu_1)\left[\pm i\pm\left(\frac{-\nu_2}{(1+\nu_1)^2}\right)^{\frac{1}{2}}\right]$$

For $|\mu| < \delta$, $\nu_2 > 0$ the eigenvalues are purely imaginary and may become rationally dependent. The resonances that occur will be close to the 1 : -1 resonance if δ is small. Define $m_\delta \in \mathbb{N}$ to be the maximal number such that for all ν with $0 < |\nu| < \delta$ we have $\mathrm{kerad}_k(S) \subset \mathrm{kerad}_k(H_2^\nu)$ for $k < m_\delta$. This means that all resonances which occur when $0 < |\nu| < \delta$ have resonance terms of degree greater than m_δ. By definition each H_2^ν-normal form up to order m_δ for F_ν $(0 < |\nu| < \delta)$ is also an S-normal form. This gives:

5.1. LEMMA. Let $F_\mu : \mathbb{R}^4 \to \mathbb{R}$ be a Hamiltonian function such that the system $(\mathbb{R}^4,\omega,F_0)$ is at nonsemisimple 1 : -1 resonance. Then there exists a parameter dependent symplectic diffeomorphism such that F_μ is transformed to H_ν, where, provided $|\nu| < \delta$, H_ν is S-invariant up to order m_δ and $H_2^\nu(x,y) = (\alpha+\nu_1)(x_1y_2-x_2y_1) + \frac{1}{2}(x_1^2+x_2^2) + \frac{1}{2}\,\nu_2(y_1^2+y_2^2)$.

In the following two sections we will show how to find an S-invariant function E_ν equal to H_ν up to order m_δ, which has the property that the periodic solutions of $(\mathbb{R}^4,\omega,E_\nu)$ correspond to those of $(\mathbb{R}^4,\omega,H_\nu)$ by means of a parameter dependent C^k-diffeomorphism, where $k = m_\delta$.

2. The Moser-Weinstein reduction

Consider the Hamiltonian system $(\mathbb{R}^4,\omega,H_\nu)$ with H_ν as in lemma 5.1. By scaling $(x,y) = (\varepsilon u,\varepsilon v)$ we get a new Hamiltonian $\hat{H}_\nu(u,v,\varepsilon) = \varepsilon^{-2} H_\nu(\varepsilon u,\varepsilon v)$. Now write $\hat{H}_\nu(u,v,\varepsilon) = \hat{H}(u,v,\nu,\varepsilon)$ and introduce a time scaling $(u(\tau t),v(\tau t)) = (\xi(t),\eta(t))$. Then the Hamiltonian function becomes

$$\hat{H}(\xi,\eta,\nu,\tau,\varepsilon) = \tau\hat{H}(\xi,\eta,\nu,\varepsilon).$$

Note that a periodic solution of $X_{\hat{H}}(\xi,\eta,\nu,\tau,\varepsilon)$ of period one corresponds to a periodic solution of X_H of period τ.

Let Per denote the space of periodic curves $\gamma : \mathbb{R}/\mathbb{Z} \to \mathbb{R}^4$ Furthermore let $\mathrm{Per}^k = \mathrm{Per} \cap C^k(\mathbb{R}/\mathbb{Z},\mathbb{R}^4)$. On Per^1 we define the vector field $\tilde{X}_{\hat{H}}(\gamma,\nu,\tau,\varepsilon)$ by

$$(5.2) \qquad \tilde{X}_{\hat{H}}(\gamma,\nu,\tau,\varepsilon)(t) = \frac{d\gamma(t)}{dt} - X_{\hat{H}}(\gamma(t),\nu,\tau,\varepsilon)$$

By construction a stationary point γ for $\tilde{X}_{\hat{H}}(\gamma,\nu,\tau,\varepsilon)$ is a periodic solution of $X_{\hat{H}}$. If we introduce on Per^1 the symplectic form $\tilde{\omega}$ defined by

$$\tilde{\omega}(\delta\gamma,\delta\gamma') = \int_{\mathbb{R}/\mathbb{Z}} \omega(\gamma(t),\gamma'(t))dt$$

with $\delta\gamma(t)(\delta\gamma'(t))$ an infinitesimal deformation of the periodic function $\gamma(\gamma')$, then the Hamiltonian function corresponding to $\tilde{X}_{\hat{H}}$ is just the action integral for $\hat{H}$ given by

$$(5.3) \qquad \tilde{H}(\gamma) = \int_{\mathbb{R}/\mathbb{Z}} [\langle\frac{d\xi}{dt}(t),\eta(t)\rangle - \hat{H}(\xi(t),\eta(t),\nu,\tau,\varepsilon)]dt$$

where $\gamma(t) = (\xi(t),\eta(t))$. Thus on Per^1 we have $X_{\tilde{H}} = \tilde{X}_{\hat{H}}$.

Consider $\alpha S(x,y)$. Each solution of the linear Hamiltonian vector field corresponding to this function is periodic with period $2\pi/\alpha$. (with exception of the origin, which is a stationary point.) Let $\tau_0 = 2\pi/\alpha$. Then $\hat{S}(\xi,\eta,\tau_0) = \tau_0\ S(\xi,\eta)$. $\hat{S}$ is the semisimple part of $\hat{H}_2(\xi,\eta,0,\tau_0,\varepsilon)$ and each solution of $X_{\hat{S}}$ has period one.

Let A_S, A_X, A_Y be the infinitesimal symplectic matrices associated to the Hamiltonian functions $\alpha S(x,y)$, $X(x,y)$ and $Y(x,y)$ respectively. We have $A_SA_X = A_XA_S$, $A_SA_Y = A_YA_S$; $\ker A_X = \operatorname{im} A_X$; $\ker A_Y = \operatorname{im} A_Y$, that is, $A_X^2 = A_Y^2 = 0$. Furthermore $\ker A_X \oplus \ker A_Y = \mathbb{R}^4$. Notice that the time 1 map of the flow of $X_{\hat{S}}$, that is, $\exp(\tau_0 A_S)$, is just the identity map on $\mathbb{R}^4$. $A = A_S + A_X$ is the matrix of the linearized system.

Let $N = \{\gamma \in Per \mid \gamma(t)=\exp(\tau_0 t A_S)z \text{ with } z \in \mathbb{R}^4\}$ and $B^k = \{\gamma \in Per^k \mid \int_0^1 \exp((1-s)\tau_0 A_S)\gamma(s)ds = 0\}$. It is easily verified that $Per^k = N \oplus B^k$. Furthermore let $\tilde{B} = B^0 + \mathbb{R}.A\gamma_0$ with $\gamma_0 \in N$.

5.4. LEMMA. $\frac{\partial \tilde{X}_{\hat{H}}}{\partial(\gamma,\tau)}(\gamma_o,0,\tau_o,0) : B^1 \times \mathbb{R} \to \tilde{B}$, $\gamma_0 \in N$, is bijective.

Proof Write $D(\gamma_0) = \frac{\partial \tilde{X}_{\hat{H}}}{\partial(\gamma,\tau_o)}(\gamma_o,0,\tau_o,0)$. Then $D(\gamma_0):Per^1 \times \mathbb{R} \to Per^0$ is given by

$$(5.5) \qquad D(\gamma_0)(\gamma,\tau)(t) = \frac{d\gamma}{dt}(t) - \tau_0 A\gamma(t) - \tau A\gamma_0(t)$$

where A is the matrix associated to $X_{H_2^\nu}$ for $\nu = 0$. For each $\zeta \in Per^0$, $\tau \in \mathbb{R}$ we may solve $D(\gamma_0)(\gamma,\tau) = \zeta$ getting

$$(5.6) \qquad \gamma(t) = \exp(t\tau_0 A)\gamma(0) + \int_0^t \exp((t-s)\tau_0 A)(\tau A\gamma_0(s) + \zeta(s))ds$$

Define $N_1 \subset N$ by $N_1 = \{\gamma \in Per | \gamma(t) = \exp(\tau_0 tA_S)z$ with $z \in \ker A_X\}$ and $N_2 \subset N$ by $N_2 = \{\gamma \in Per | \gamma(t) = \exp(\tau_0 tA_S)z$ with $z \in \ker A_Y\}$. Then it follows from $\ker A_X \oplus \ker A_Y = \mathbb{R}^4$ that $N_1 \oplus N_2 = N$. Thus we have

$$(5.7) \qquad Per^1 \times \mathbb{R} = N_1 \times \{0\} \oplus N_2 \times \{0\} \oplus B^1 \times \{0\} \oplus \{0\} \times \mathbb{R}.$$

Moreover, $D(\gamma_0)(N_1 \times \{0\}) = 0$, $D(\gamma_0)(N_2 \times \{0\}) \subset N_1$, $D(\gamma_0)(B^1 \times \{0\}) \subset B^0$ and $D(\gamma_0)(\{0\} \times \mathbb{R}) = \mathbb{R}.A\gamma_0$, which follows easily using (5.5). If $\gamma_0 \in N_1$ then $D(\gamma_0)(N_2 \times \{0\}) \cap D(\gamma_0)(\{0\} \times \mathbb{R}) \neq \{0\}$ which means that there are elements in $N_2 \times \mathbb{R}$ which belong to the kernel of $D(\gamma_0)$. It is clear that each nonzero element in $\ker(D(\gamma_0)) \subset N \times \mathbb{R}$ has nonzero N-component. Thus $\ker D(\gamma_0) \cap (B^1 \times \mathbb{R}) = \{(0,0)\}$. For $\gamma_0 \notin N_1$ it is obvious that $\ker D(\gamma_0) = N_1 \times \{0\}$ and thus $\ker D(\gamma_0) \cap (B^1 \times \mathbb{R}) = \{(0,0)\}$. Therefore for each $\gamma_0 \in N$, $D(\gamma_0) : B^1 \times \mathbb{R} \to \tilde{B}$ is injective.

For each element of $\tilde{B}$ the inverse can be read off from (5.6), which means that $D(\gamma_0) : B^1 \times \mathbb{R} \to \tilde{B}$ is also surjective. Thus $D(\gamma_0) : B^1 \times \mathbb{R} \to \tilde{B}$ is bijective □

Let $\tilde{N}$ be the orthogonal complement of $\mathbb{R}.A\gamma_0$ in N with respect to some suitable inner product. By averaging we can always take this inner product to be S-invariant. In this case because S is in normal form we may just take the standard inner product. Then $\tilde{N} \oplus \tilde{B} = Per^0$

and we can write $\tilde{X}_{\hat{H}}(\gamma,\nu,\tau,\varepsilon) = \tilde{X}_{\hat{H}}(\gamma,\nu,\tau,\varepsilon)_{\tilde{N}} + \tilde{X}_{\hat{H}}(\gamma,\nu,\tau,\varepsilon)_{\tilde{B}}$.

5.8. LEMMA. For each $\gamma_0 \in N$ there exist neighborhoods V_1, V_2 of zero in $\mathbb{R}$, a neighborhood U_1 of zero in B^1 and a neighborhood U_2 of τ_0 in $\mathbb{R}$ such that for $\varepsilon \in V_1$, $\nu \in V_2$ $\tilde{X}_{\hat{H}}(\gamma_0+\gamma_B,\nu,\tau,\varepsilon)_{\tilde{B}} = 0$ has a unique solution $(\gamma_B(\gamma_0,\nu,\varepsilon), \tau(\gamma_0,\nu,\varepsilon)) \in U_1 \times U_2$ depending smoothly on γ_0,ν,ε.

proof. $\tilde{X}_{\hat{H}}(\gamma_o,0,\tau_o,0)_{\tilde{B}} = 0$. By lemma 5.4. we may apply the implicit function theorem to obtain γ_B,τ as smooth functions of γ_0,ν,ε. ⊠

Recall that $Per^0 = N \oplus B^0 = \tilde{N} \oplus \mathbb{R}.A\gamma_0 \oplus B^0 = \tilde{N} \oplus \tilde{B}$. If we consider the equation.

$$(5.9) \qquad \tilde{X}_{\hat{H}}(\gamma,\nu,\tau,\varepsilon) = 0$$

then by lemma 5.8. we may solve the $\tilde{B}$-part of this equation for τ and the B^1-part γ_B of γ. Substituting $\gamma_B(\gamma_0,\nu,\varepsilon)$ and $\tau(\gamma_0,\nu,\varepsilon)$ into (5.9) (writing γ_B for $\gamma_B(\gamma_0,\nu,\varepsilon)$ and τ for $\tau(\gamma_0,\nu,\varepsilon)$) we obtain $\tilde{X}_{\hat{H}}(\gamma_0+\gamma_B,\nu,\tau,\varepsilon)=0$ with $\tilde{X}_{\hat{H}}(\gamma_0+\gamma_B,\nu,\tau,\varepsilon) \in \tilde{N}$ because the $\tilde{B}$-part vanishes. Thus $\tilde{X}_{\hat{H}}(\gamma_0+\gamma_B,\nu,\tau,\varepsilon)_N + \tilde{X}_{\hat{H}}(\gamma_0+\gamma_B,\nu,\tau,\varepsilon)_B = \tilde{X}_{\hat{H}}(\gamma_0+\gamma_B,\nu,\tau,\varepsilon) \in \tilde{N}$, therefore the B-part is zero. Thus $\tilde{X}_{\hat{H}}(\gamma_0+\gamma_B,\nu,\tau,\varepsilon) = 0$ is equivalent to

$$(5.10) \qquad \tilde{X}_{\hat{H}}(\gamma_0+\gamma_B,\nu,\tau,\varepsilon)_N = 0.$$

By definition we may write (5.10) as $\frac{d\gamma_0}{dt} = \tau\, X_{\hat{H}}(\gamma_0+\gamma_B,\nu,\varepsilon)_N$ using the fact that $\frac{d\gamma_B}{dt} \in B$. Because $\frac{d\gamma_0}{dt} = X_{\hat{S}}(\gamma_0)$ we get

$$(5.11) \qquad X_{\hat{S}}(\gamma_0) = \tau\, X_{\hat{H}}(\gamma_0+\gamma_B,\nu,\varepsilon)_N .$$

Now consider the action of e^{sA_S}. Then $e^{sA_S}\gamma_0(t) = \gamma_0(t+s)$. Furthermore $e^{sA_S}\tau(\gamma_0,\nu,\varepsilon) = \tau(e^{sA_S}\gamma_0,\nu,\varepsilon)$ and $e^{sA_S}\gamma_B(\gamma_0,\nu,\varepsilon) = \gamma_B(e^{sA_S}\gamma_0,\nu,\varepsilon)$ because γ_B and τ are locally unique solutions by lemma 5.8.

If follows that $X_{\hat{H}}(\gamma_0+\gamma_B,\nu,\varepsilon)_N$ is invariant under the action of the flow of X_S on Per. Let $\hat{E}(\gamma_0(t),\nu,\varepsilon) = \hat{H}(\gamma_0+\gamma_B,\nu,\varepsilon)$. Then E is S-invariant. Furthermore we may write (5.11) as $d\hat{S} = \tau d\hat{E}$. By the identification $\gamma_0 \to \gamma_0(0)$ we obtain a function $\hat{E}(\xi,\eta,\nu,\varepsilon)$ on $\mathbb{R}^4$. The S-action on Per reduces to the S-action on $\mathbb{R}^4$. Thus we obtain the following equation

(5.12) $$dS(\xi,\eta) = \tau d\hat{E}(\xi,\eta,\nu,\varepsilon)$$

which is equivalent to (5.11) and (5.10). Any solution $(\xi,\eta,\tau(\xi,\eta,\nu,\varepsilon))$ of (5.12) corresponds to a solution $\gamma_0 + \gamma_B$, $\tau(\gamma_0,\nu,\varepsilon)$ of (5.9) and thus to a periodic solution $\gamma_0 + \gamma_B(\gamma_0,\nu,\varepsilon)$ of period $\tau(\gamma_0,\nu,\varepsilon)$ of $X_{\hat{H}}$ (Here $\gamma_0 = e^{\tau_0 t A_S}(\xi,\eta)$). Thus we have proved the following lemma.

5.13. <u>LEMMA</u>. For ε,ν sufficiently small there is a S-invariant C^∞ function $\hat{E}(\xi,\eta,\nu,\varepsilon)$, depending C^∞ on ε and ν such that the set of points, where $d\hat{E}(\xi,\eta,\nu,\varepsilon)$ is a multiple τ of $dS(\xi,\eta)$, corresponds to the set of periodic solutions of $(\mathbb{R}^4,\omega,\hat{H}(\xi,\eta,\nu,\varepsilon))$ by the C^∞-map $(\xi,\eta) \to (\xi,\eta) + \gamma_B(\xi,\eta,\nu,\varepsilon)$. The period of these solutions is equal to τ^{-1}.

5.14. <u>REMARK</u>. In the above lemma we are considering the scaled vector field. The fact that the scaling is singular at the origin influences the smoothness after rescaling. This will be studied in the next section.

3. Differentiability of E(x,y,ν)

Suppose that $H(x,y,\nu)$ is in S-normal form up to order m. Then it is obvious that $\tilde{X}_{\hat{H}}(\gamma_0,\tau,\nu,\varepsilon) \in N$ modulo $O(\varepsilon^m)$. Recall that $\tilde{N}$ is the orthogonal complement to $\mathbb{R}.A_{\gamma_0}$ in N with respect to a suitable S-invariant inner product. Denote this inner product by $\langle .,. \rangle$. Then we may solve $\langle A\gamma_o,\tilde{X}_{\hat{H}}(\gamma_o,\nu,\tau,\varepsilon)\rangle = 0$ for τ getting $\bar{\tau}(\gamma_0,\nu,\varepsilon)$. It follows that

(5.15) $$\tilde{X}_{\hat{H}}(\gamma_o,\bar{\tau},\nu,\varepsilon)_{\tilde{B}} = O(\varepsilon^m)$$

From (5.15) and lemma 5.8. we find that $\gamma_B(\gamma_0,\nu,\varepsilon) = O(\varepsilon^m)$ and $\tau(\gamma_0,\nu,\varepsilon) = \bar{\tau}(\gamma_0,\nu,\varepsilon) + O(\varepsilon^m)$.

Recall that $\hat{H}(\gamma_0,\nu,\tau,\varepsilon) = \varepsilon^{-2} H(\varepsilon\gamma_0,\nu,\tau)$. Running through the reduction gives $\gamma_B(\gamma_0,\nu,\varepsilon) = \varepsilon^{-1} \gamma_B(\varepsilon\gamma_0,\nu)$, where $\gamma_B(\varepsilon\gamma_0,\nu)$ is a function not depending on ε. Now $\hat{E}(\gamma_0,\nu,\varepsilon)$ is the Hamiltonian corresponding to $\hat{X}_{\hat{H}}(\gamma_0+\gamma_B,\nu,\varepsilon)_N = \varepsilon^{-1} \hat{X}_{\hat{H}}(\varepsilon\gamma_0+\gamma_B(\varepsilon\gamma_0,\nu),\nu)_N$. Thus $\hat{E}(\gamma_0,\nu,\varepsilon) = \varepsilon^{-2} E(\varepsilon\gamma_0,\nu)$. By the identification $\gamma_0 \to \gamma_0(0)$ we may consider, $\hat{E}$ and E as functions on $\mathbb{R}^4$, where we have the scaling $(x,y) = (\varepsilon\xi,\varepsilon\eta)$. This gives

$$(5.16) \qquad \gamma_B(x,y,\nu) = \varepsilon.\gamma_B(\varepsilon^{-1}\xi,\varepsilon^{-1}\eta,\nu,\varepsilon)$$

To analyze the smoothness of E it is sufficient to consider the right hand side of (5.16). Because $\gamma_B(\gamma_0,\nu,\varepsilon) = O(\varepsilon^m)$ we find that the derivatives of $\gamma_B(x,y,\nu)$ with respect to (x,y) vanish up to order $m - 1$ at $(x,y) = (0,0)$. Thus $E(x,y,\nu)$ is at least C^m over $(0,0)$. Notice that on $U^* = \{0 < |(x,y)| < d\}$, $E(x,y,\nu)$ is C^∞.

Furthermore we have that $E(x,y,\nu) = H(x,y,\nu) + O(|(x,y)|^{m+1})$ that is, E equals the S-normal form of H up to order m. We may now rephrase lemma (5.14) in terms of functions on $\mathbb{R}^4$ not depending on ε.

5.17. <u>THEOREM</u>. For ν sufficiently small there is a S-invariant function $E(x,y,\nu)$ with the following properties:

a)up to order m $E(x,y,\nu)$ is equal to the S-normal form of $H(x,y,\nu)$ (H can be normalized up to arbitrary order, provided ν is sufficiently small, see section 1).

b)On U^* E is a C^∞ function depending C^∞ on ν.

c)On $U = \{(x,y) \in \mathbb{R}^4 \mid |(x,y)| < d\}$ E is a C^m function depending C^∞ on ν.

d) The set of points where $dE(x,y,\nu)$ is a multiple of $dS(x,y)$ corresponds to the set of periodic solutions of the system $(\mathbb{R}^4, \omega, H_\nu)$ by a map $(x,y) \to (x,y) + \gamma_B(x,y,\nu)$ which is C^∞ on U^* and C^{m+1} on U for U^* and U sufficiently small. The period is the reciprocal of the multiple and is close to ω_o.

Recall that the set of points where dE is a multiple of dS is just the critical point set $\Sigma_{E\times S}$ of the energy-momentum map $E \times S$. This allows us to combine theorem 5.17. and the theory of chapter 3 to obtain the main theorem of this thesis.

5.18. THEOREM. Suppose we have a system $(\mathbb{R}^4, \omega, H_\mu)$ depending C^∞ on a parameter μ (possibly a vector) which for $\mu = \mu_o$ is in $1 : -1$ nonsemi-simple resonance (that is, H_{2,μ_o} equals (3.24)). Furthermore suppose that in the S-normal form of H_μ for $\mu = \mu_o$ the coefficient of $Y^2 = (y_1^2+y_2^2)^2$ is nonzero. Then there exists a C^∞ map $\mu \to \nu(\mu)$, $\nu(\mu_o) = 0$ and a parameter dependent map φ_μ (depending C^∞ on μ) such that for μ close to μ_o the set of periodic orbits of $(\mathbb{R}^4, \omega, H\ )$ near the origin and with period close to $\tau_o = 1/\alpha$ is equal to the φ_μ-image of the set $\Sigma_{G_\nu \times S}$ of points where dG_ν is a multiple of dS. By the result of section 1 for μ sufficiently close to μ_o, φ_μ can be made to be C^m for arbitrary m with respect to the phase space variables.

5.19. REMARK. Notice that instead of the theory of chapter 3 we in fact need a C^k-theory. However the main ingredients of this theory such as the C^k-preparation theorem do exist (Lasalle [1973], Vegter [1981]). Furthermore because of the remarks made in section 1 we may start with a function which is differentiable of arbitrary degree. Therefore we can always deal with the loss of differentiability in the process and obtain the same results as in chapter 3.

5.20. <u>REMARK</u>. Starting with a general Hamiltonian H in 1 : -1 nonsemisimple resonance it follows from normal form theory that the coefficient a of Y^2 in the normal form of H is a polynomial function of the coefficients of H_2, H_3 and H_4 in H. Thus in the space of coefficients of the original Hamiltonian the equation a = 0 determines an algebraic variety. In theorem 5.18 we restrict ourselves to the complement of this variety. Such a complement is open and dense. Thus 5.18. describes the generic case.

5.21. <u>REMARK</u>. One might expect a reduction to the 2-dimensional space of periodic solutions of the linearized system, instead of a reduction to the 4-dimensional space of periodic solutions of the semisimple part of the linearized system. However, the variety of periodic solutions of the system in normal form (as described in section 3 chapter 4) does not fit into a smooth manifold containing the origin of dimension smaller than four. So the weaker reduction performed here is essential.

4. Invariant tori, homoclinic orbits

From the preceding sections it is clear that the relative equilibria of the integrable system (considered in chapter 4) survive non integrable perturbations. The next question is what other features of the integrable system or family of systems persist under perturbation.

Two features are of particular interest: first the foliation into invariant tori and second the homoclinic orbits in the hyperbolic case a > 0 when stable and unstable manifold coincide.

In chapter 4 above large parts of the energy-momentum plane we found a foliation of phase space into invariant tori. If one does consider the complete family of systems one finds a foliation into tori above large parts of the energy-momentum-detuning space. For each

ν = constant, $\nu > 0$ section one has the classically considered cases of systems with purely imaginary eigenvalues which are not in resonance or are in a high-order resonance. For these systems Pöschel [1982] proves that the perturbed system possesses smooth invariant tori for a Cantor set of frequencies. On sections ν = constant, $\nu \leq 0$ the existence of invariant tori is still open, although considering the geometric equivalence with the situation for $\nu > 0$, one easily jumps to the conjecture that also in these cases the same conclusions hold. For the case $\nu > 0$ one has to deal with the stable and unstable manifold and with the problem of monodromy. These questions come together in studying the foliation of invariant tori over a neighborhood of the origin in (G,S,ν)-space. Here the problem is to prove the existence of quasi periodic solutions for a one parameter family of perturbed integrable systems.

So far we considered the question of existence of quasiperiodic solutions near the origin. Another question is the existence of quasi-periodic solutions near the perturbed elliptic relative equilibria.

Another interesting point found in chapter 4 is that for the integrable system for $\nu < 0$, $a > 0$ (the hyperbolic equilibrium case). the stable and unstable manifold coincide. The question is then which homoclinic orbits persist under perturbation. In general we can make the remark (see Robinson [1970]) that transversal intersection of stable and unstable manifolds is generic for Hamiltonian systems. Also here we may pose similar questions concerning the homoclinic behaviour near the hyperbolic relative equilibria.

Chapter VI

The restricted problem of three bodies

0. Introduction

In this final chapter it will be shown how the restricted three body problem fits into the theory developed in the previous chapters. First (section 1) we will show that the equations of motion of the restricted problem of three bodies at the equilateral equilibrium point L_4 and for Routh critical mass ratio are of the type considered in the foregoing chapters. In section 2 we will give some historical information about the restricted three body problem which is over two centuries old. During these two centuries many mathematicians, physicists and astronomers have been occupied with the many aspects of the problem for all kinds of reasons. We give a review of the papers concerning the non-semisimple 1:-1 resonance and discuss their relation with our results.

For a more complete treatment of all aspects of the restricted problem of three bodies, including an extensive bibliography, we refer the reader to Szebehely [1967].

1. The equations of motion of the restricted three body problem

The restricted three body problem can be described as follows: two bodies, called the primaries P_1 and P_2, move in circular orbits around their common center of mass according to the laws of gravity of Newton while a third body P with negligible mass moves in the same plane. The problem is to describe the motion of the body P which is acted upon by P_1 and P_2 according to Newton's laws but does not perturb the motion of P_1 and P_2 because it is supposed to have no mass.

Sometimes this problem is called the planar circular restricted problem of three bodies. This name refers to all the restrictions made when compared with the general problem of three bodies: three bodies moving in three dimensional space according to Newton's laws. Planar indicates that all bodies move in the same plane, circular stands for the circular movement of the primaries, and restricted stands for the fact that P is supposed to be without mass.

We fix the units of length, mass, and time as follows. As unit of length we choose the distance between the two finite masses P_1 and P_2, as the unit of mass the sum of the masses of P_1 and P_2, and as the unit of time we choose the angular velocity of P_1 and P_2. With this choice the gravitational constant becomes one. The motion of P will be referred to in a co-ordinate frame whose origin is the center of mass of P_1 and P_2 and which rotates so that P_1 and P_2 are on the horizontal axis with P_2 having positive coordinates.

fig. (6.1)

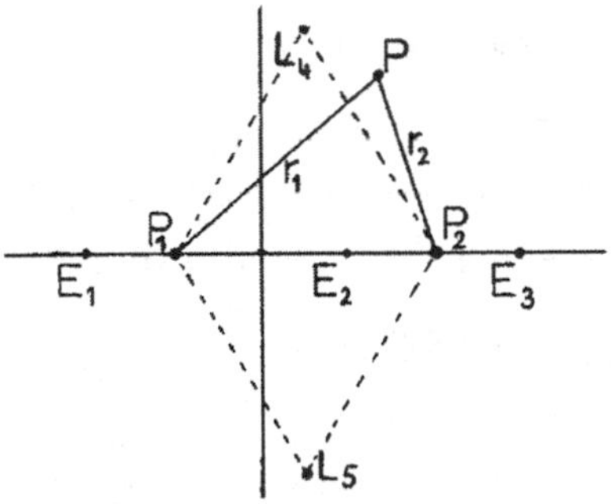

If we suppose P_1 to have the larger mass we may call its value $1 - \mu$, P_2 consequently having mass μ, with $0 < \mu \leq \frac{1}{2}$. Then the coordinates of P_1 and P_2 are $(0,-\mu)$ and $(0,1-\mu)$ resp. (see fig. (6.1)).

The equations of motion can be put in Hamiltonian form with Hamiltonian function

$$(6.1) \qquad H(x,y) = \tfrac{1}{2}(y_1^2+y_2^2) - (x_1y_2-x_2y_1) - \frac{1-\mu}{r_1} - \frac{\mu}{r_2}$$

Here (x_1,x_2) are the coordinates of P in the plane, (y_1,y_2) are the corresponding momenta, r_1 and r_2 are the distances of P to P_1 and P_2, that is,

$$r_1^2 = (x_1+\mu)^2 + x_2^2 \qquad r_2^2 = (x_1 - (1-\mu))^2 + x_2^2$$

With ω the standard symplectic form and H as in (6.1) we obtain a Hamiltonian system $(\mathbb{R}^4,\omega,H)$. X_H has five stationary points. The points $E_{1,2,3}$ called the Euler equilibrium points or collinear equilibrium points. These points were first described by Euler [1765]. In addition to $E_{1,2,3}$ we have $L_{4,5}$, the equilateral or Lagrange equilibrium points, which were first computed by Lagrange [1772]. The Lagrange points form an equilateral triangle with P_1 and P_2. (see fig. (6.1)).

We are interested in the motion of P near the point L_4 having coordinates $(\frac{1}{2}(1-2\mu),\frac{1}{2}\sqrt{3})$. This also gives us the motion near L_5 because of the symmetry with respect to the horizontal axis. A translation makes L_4 the origin of our co-ordinate system. An additional rotation over an angle β, with $\tan 2\beta = \sqrt{3}(1-2\mu)$, finally gives us the Hamiltonian system with Hamiltonian function $H(\xi,\eta)$

$$(6.2) \qquad H(\xi,\eta) = H_2(\xi,\eta) + \sum_{p=0}^{\infty}\sum_{q=0}^{\infty} \omega_{pq}\, \xi_1^p \xi_2^q$$

where

$$(6.3) \qquad H_2(\xi,\eta) = \tfrac{1}{2}(\eta_1^2+\eta_2^2) - (\xi_1\eta_2-\xi_2\eta_1) - \tfrac{1}{4}(1-\tfrac{3}{2}\delta)\xi_1^2 - \tfrac{1}{4}(1+\tfrac{3}{2}\delta)\xi_2^2$$

with $\delta^2 = 1 + 3(1-2\mu)^2$, see Deprit [1966[a]]. The matrix of the linearized system is

$$A = \begin{pmatrix} 0 & 1 & 1 & 0 \\ -1 & 0 & 0 & 1 \\ \frac{1}{2}(1-\frac{3}{2}\delta) & 0 & 0 & 1 \\ 0 & \frac{1}{2}(1+\frac{3}{2}\delta) & -1 & 0 \end{pmatrix}$$

The characteristic equation of A is $\lambda^4 + \lambda^2 + \mu\frac{27}{4}(1-\mu) = 0$. For $\mu = \mu_0 = \frac{1}{2}(1-\frac{1}{9}\sqrt{69})$ we find the eigenvalues $\pm i\frac{1}{2}\sqrt{2}$, $\pm i\frac{1}{2}\sqrt{2}$. For $0 < \mu < \mu_0$ we find four purely imaginary eigenvalues which are all different. For $\mu_0 < \mu \leq \frac{1}{2}$ we find four different eigenvalues all having nonzero real and imaginary parts. In fact for μ passing through μ_0 the eigenvalues behave as described in fig. (1.2) ch.I sect.3.

In Burgoyne and Cushman [1974] a linear change of coordinates is found which transformes the Hamiltonian vector field X_{H_2} to its linear normal form, that is, in the new coordinates H_2 becomes:

$$H_2(x,y) = \tfrac{1}{2}\sqrt{2}\,(x_1y_2-x_2y_1) + \tfrac{1}{2}(x_1^2+x_2^2)$$

In van der Meer [1982] it was shown how to transform the higher order terms to their normal form. Up to order four we have

$$H(x,y) = \tfrac{1}{2}\sqrt{2}(x_1y_2-x_2y_1) + \tfrac{1}{2}(x_1^2+x_2^2)$$
$$+ a(y_1^2+y_2^2)^2 + b(y_1^2+y_2^2)(x_1y_2-x_2y_1) + c(x_1y_2-x_2y_1)^2$$

with $a > 0$ (i.e. $a = 59/864$).

This shows that the restricted three body problem at L_4 for $\mu = \mu_0$ is in nonsemisimple 1 : -1 resonance. μ_0 is called the critical mass value of Routh. Furthermore for μ passing through μ_0 we have a Hamiltonian Hopf bifurcation. Because the coefficient of $(y_1^2+y_2^2)^2$ in the normal form is positive, the behaviour of periodic solutions is, up to C^k-diffeomorphism, as described in chapter 4 for $a > 0$ (see fig. 4.12).

2. History of the problem

The formulation of the general three body problem, made possible by the theories of Kepler and Newton, goes back to Euler who stated the problem as early as 1727 in his diary (see Volk [1976]). In his paper "Considerationes de motu corporum coelestium" (1764) Euler says that the general three body problem is to hard to solve. Therefore he restricts to "problema restricta" such as a Sun-Earth-Moon problem with the Moon having zero mass and being in conjunction or opposition to Sun-Earth. (see Volk [1983]) According to Szebehely [1967] and others the restricted problem as we know it, was formulated by Euler [1772]. The collinear problem is formulated in Euler [1765]. It is here that the existence of the collinear libration points is established. These points were rediscovered by Lagrange [1772] who treated a more general problem and found also the equilateral equilibria. Also Jacobi [1836] studied the restricted three body problem. Some people give him the credit for its first formulation.

In those days the main interest was in the Sun-Earth-Moon problem because of the importance for navigation and geography. Already Euler saw the three body problem as a pure mathematical model problem with other astronomical applications. Beside Euler, Lagrange and Jacobi of course many others were occupied with these problems.

Poincaré, another great name in the history of the three body problem, used this problem as the main example to illustrate his more general ideas. In his works the name restricted three body problem is formally introduced for the problem described in section 1. In volume 1 page 11 of his "Méthodes Nouvelles" he speaks of "cas particuliers du Problème des trois Corps" but in volume III page 69 he formally introduces "le problème restreint". In his time and certainly in his prize memoir (Poincaré [1890]) the three body problem was an example in considerations concerning the stability of the solar system.

In 1884 Gyldén began the study of infinitesimal orbits near the equilateral equilibria. At the beginning of this century this became of importance when in 1906 Achilles, the first of the Trojan satellites, was discovered to be near to the Lagrange equilibrium of the Sun-Jupiter system. Later this century the restricted three body problem became again of interest in problems concerning artificial satellites.

Since the beginning of this century a number of people have considered periodic solutions near the equilateral equilibria for all mass ratios including the particular resonance at the critical mass ratio of Routh considered in these notes. In Routh [1875] this critical mass ratio is computed for the first time as an exceptional value among the mass ratios

The first person, to my knowledge who studied the passage through resonance was Brown [1911]. Although he considers a slightly different model placing the larger body in the origin with the other primary rotating about it, the equations he obtained are analogous to ours. He considers the series expansions for the solutions up to third order. More precisely he considers the Hamiltonian function up to fourth order, if we suppose that the solutions formally are given by exp(t ad(H)). He found that for $\mu > \mu_0$ the family of periodic solutions is no longer attached to the origin, but to some limiting orbit (which in our language is the relative equilibrium for $S = 0$) lying on some distance from the origin. That his analysis was sufficient to describe the behaviour of these families of periodic solutions is obvious now we know that the Hamiltonian is 4-determined.

In Pedersen [1934] we find a more detailed description of the behaviour near $\mu = \mu_0$. He uses the same model as described in section 1 and again gives a third order analysis of the solutions. He adds nothing new to the results of Brown.

In Buchanan [1939] we find a complex Jordan normal form for the linearized system for $\mu = \mu_0$. In Buchanan [1941] the families of periodic solutions at $\mu = \mu_0$ are computed.

In Deprit [1966^b] we find a fourteenth order analysis of the limiting orbits for $\mu > \mu_0$. In Deprit and Henrard [1968], [1969] an extensive treatment of the behaviour of periodic solutions in the restricted three body problem is given, including what is known up until then about the behaviour at the critical mass ratio of Routh. In the second paper an almost complete description of the behaviour of periodic solutions during the passage through resonance is given as a conjecture based on numerical results.

The problem is again taken up by Meyer and Schmidt [1970] who actually prove the existence of families of periodic solutions at L_4 for mass ratios near Routh critical value. Their paper is followed by a paper of Schmidt and Sweet [1973] who treat the problem at resonance in a different way. Yet another way is given in Roels [1975]. The three last metioned papers all use complex normal forms which has the disadvantage mentioned in Burgoyne and Cushman [1974]. In Meyer and Schmidt [1970] and Schmidt and Sweet [1973] the scaling $x \to \varepsilon^2 x$, $y \to \varepsilon y$, $\nu \to \varepsilon^2 \nu$ is used. Such a scaling has the disadvantage that in a neighborhood of the origin one restricts to a cusplike area having contact of order two in y-direction. It is a priori not clear that the phenomena to be studied take place in this area. In our terms this scaling gives $G \to \varepsilon^4 G$, $S \to \varepsilon^3 S$ and $\nu \to \varepsilon^2 \nu$. Using the results of chapter 4 it is now easily checked that in scaled variables the following conclusions can be drawn: For $a > 0$, $\nu > 0$ one finds two families of periodic orbits. Because one has to stay away from the G,ν-plane, one can only conclude that in the limit $S \to 0^{\pm}$ these families are attached to the origin. For $a > 0$, $\nu < 0$ these families persist and do not contain the origin. All periodic orbits are elliptic. For $a < 0$, $\nu > 0$ again two families are found which do contain a transitional orbit where the stability type changes from elliptic to hyperbolic. In the limit $S \to 0^{\pm}$ the hyperbolic families are connected and elliptic families are attached to the origin. For $\nu \to 0$ the families pull into the origin. These results agree with theorem (A) in Meyer and Schmidt

[1970]. What is not found in this way is how the families are attached to the origin, to each other, and to the 'organizing center' (the origin) in G,S,ν-space.

In Meyer [1974] a short survey of the behaviour of periodic solutions is given. The pictures in this article show that, although not all details of the behaviour at the passage through nonsemisimple 1 : -1 resonance were proved, people did have a good picture of what happens.

Recently Caprino, Maffei and Negrini [1984], and Dell'Antonio and D'Onofrio [1983] considered the Hamiltonian Hopf bifurcation as a special case in a more general treatment of families of periodic solutions of Hamiltonian systems. The first paper is devoted to establishing the existence of families of periodic solutions. The second paper concentrates on the number of periodic solutions.

Starting with the formulation of the restricted problem of three bodies by Euler it is clear the the astronomical and mathematical interest in this problem go hand in hand. The 'modern' mathematical point of view might be illustrated by the following quotation of Birkhoff [1915] which is still up to date.

"Thorough investigation of non-integrable dynamical problems is essential for the further progress of dynamics. Up to the present time only the periodic movements and certain closely allied movements have been treated with any degree of success in such problems, but the final goal of dynamics embraces the characterization of all types of movement, and of their interrelation.

The so-called restricted problem of three bodies, in which a particle of zero mass moves subject to the attraction of two other bodies of positive mass rotating in circles about their centre of gravity, affords a typical and important example of a non-integrable dynamical system".

The problem of non-integrability originates from Poincarés work. It has finally led to the modern problems of invariant tori, homölinic and heteroclinic behaviour. Although nowadays these topics are stüdied extensively still many questions remain open concerning non-integrable systems.

References

Abraham, R. and Marsden, J.E.: 1978, *Foundations of mechanics, 2nd. ed.* Benjamin/Cummings, Reading, Massachussetts.

Arnold, V.I.: 1971, *On matrices depending on parameters.* Russian Math. Surveys 26, 29-43. Also Arnold [1981], 46-60.

Arnold, V.I.: 1975, *Critical points of smooth functions and their normal forms.* Russian Math. Surveys 30, no. 5, 1-75. Also Arnold [1981], 132-206.

Arnold, V.I.: 1978, *Mathematical methods of classical mechanics.* G.T.M. 60, Springer Verlag, New York.

Arnold, V.I.: 1981, *Singularity theory.* London Math. Soc. Lect. Note Series 53, Cambridge University Press, Cambridge.

Bierstone, E: 1980, *The structure of orbit spaces and the singularities of equivariant mappings.* Monografias de Matematica 35, Inst. de Mat. pura et aplicada, Rio de Janeiro.

Birkhoff, G.D.: 1915, *The restricted problem of three bodies.* Rend. Circ. Math. Palermo 39, 265-334. Also Collected Papers, vol. 1, 682-751, A.M.S., New York, 1950.

Birkhoff, G.D.: 1927, *Dynamical systems.* A.M.S. coll. publications IX, New York.

Briot and Bouquet: 1856, *Recherches sur les propriétés des fonction définies par des equations differentielles.* Journal de l'École Impériale Polytechnique 21 : 36, 133-198.

Brjuno, A.D.: 1971, *Analytical form of differential equations.* Trans. Moscow Math. Soc. 25, 131-288.

Brjuno, A.D.: 1972, *The analytical form of differential equations,* II. Trans. Moscow Math. Soc. 26, 199-238.

Broer, H.W.: 1979, *Bifurcations of singularities in volume preserving vector fields.* Thesis, University of Groningen.

Broer, H.W.: 1980, *Formal normal form theorems for vector fields and some consequences for bifurcations in the volume preserving case.* In: Dynamical systems and turbulence, Warwick 1980, LNM 898, Springer Verlag, Berlin etc.

Brown, E.W.: 1911, *On the oscillating orbits about the triangular equilibrium points in the problem of three bodies.* Mon. Not. Roy. Astron. Soc. 71, 492-502.

Buchanan, D.: 1939, *A transformation to the normal form.* Rend. Circ. Matem. Palermo 62, 385-387.

Buchanan, D.: 1941, *Trojan satellites (limiting case).* Trans. Roy. Soc. Canada sect. III(3), 9-25.

Burgoyne, N. and Cushman, R.: 1974, *Normal forms for real linear Hamiltonian systems with purely imaginary eigenvalues.* Cel. Mech. 8, 435-443.

Burgoyne, N. and Cushman, R.: 1976, *Normal forms for real linear Hamiltonian systems.* In: The 1976 Ames Research Center (NASA) Conference on Geometric Control Theory, ed. C. Martin, Math. Sci. Press, Brookline, Mass., 1977.

Caprino, S.; Maffei, C. and Negrini, P.: 1984, *Hopf bifurcation at 1 : 1 resonance.* Nonlinear Analysis 8, 1011-1032.

Chen, K.T.: 1963, *Equivalence and decomposition of vector fields about an elementary critical point.* Am. J. Math. 85, 693-722.

Cherry, T.M.: 1927, *On the solution of Hamiltonian systems of differential equations in the neighborhood of a singular point.* Proc. London Math. Soc. series 2, 27, 151-170.

Churchill, R.C.; Kummer, M. and Rod, D.L.: 1983, *On averaging, reduction and symmetry in Hamiltonian systems.* J. Diff. Eq. 49, 359-414.

Cushman, R.: 1982[a], *Reduction of the nonsemisimple 1 : 1 resonance.* Hadronic J. 5, 2109-2124.

Cushman, R.: 1982[b], *Geometry and bifurcation of the normalized reduced Henon-Heiles family.* Proc. Roy. Soc. London A 382, 361-371.

Cushman, R.: 1983 , *Geometry of the energy momentum mapping of the spherical pendulum.* C.W.I. Newsletter 1, 4-18.

Cushman, R.: 1984 , *Normal form for Hamiltonian vector fields with periodic flows.* In: Differential geometric methods in mathematical physics, 125-144,ed. S.Sternberg, Reidel, Dordrecht,Holland.

Cushman, R.; Deprit, A. and Mosak, R.: 1983, *Normal form and representation theory.* J. Math. Phys. 24, 2103-2116.

Cushman, R. and Rod, D.L.: 1982, *Reduction of the semisimple 1:1 resonance.* Physica D 6, 105-112.

Dell'Antonio, G.F. and D'Onofrio, B.: 1983, *On the number of periodic solutions of a Hamiltonian system near an equilibrium point.* Preprint Inst. di Mat. Università del'Aquila, Italy.

Deprit, A.: 1966[a], *Motion in the vicinity of the triangular libration centers.* In: Lectures in Appl.Math. VI, Space Math., part 2, ed. J. Barkley Rosser, A.M.S., Providence, Rhode Island, 1966.

Deprit, A.: 1966[b], *Limiting orbits around the equilateral centers of libration.* Astron. J. 71, 77-87.

Deprit, A.: 1969, *Canonical transformations depending on a small parameter.* Cel. Mech. 1, 12-30.

Deprit, A.: 1981, *The elimination of the parallax in satellite theory.* Cel. Mech. 24, 111-153.

Deprit, A.: 1982, *Delaunay normalisations.* Cel. Mech. 26, 9-21.

Deprit, A.: 1983, *Normalizing by factoring out the rotation.* Preprint N.B.S., submitted to Cel. Mech.

Deprit, A. and Henrard, J.: 1968, *A manifold of periodic orbits.* Adv. Astron. Astroph. 6 1-124.

Deprit, A. and Henrard. J.: 1969, *The Trojan manifold-survey and conjectures.* In: Periodic orbits, stability and resonances, ed. G.E.O. Giacaglia, Reidel, Dordrecht, Holland.

Deprit, A.; Henrard, J.; Price, J.F. and Rom, A.: 1969, *Birkhoff's normalization*. Cel. Mech. 1, 225-251.

Dugas, R.: 1950, *Histoire de la mécanique*. Éditions du griffon, Neuchatel.

Duistermaat, J.J.: 1980, *On global action angle coordinates*. Comm. Pure Appl. Math. 33, 687-706.

Duistermaat, J.J.: 1981, *Periodic solutions near equilibrium points of Hamiltonian systems*. In: Analytical and numerical approaches to assymtotic problems in analysis (Proc. Conf. Univ. Nijmegen, Nijmegen, 1980), 27-33, ed. O. Axelsson et al., North Holland Math. Studies 47, North Holland, Amsterdam, 1981.

Duistermaat, J.J.: 1983[a], *Non integrability of the 1 : 1 : 2 resonance*. Preprint University of Utrecht 281. Accepted for publication in Ergodic theory and dynamical systems.

Duistermaat, J.J.: 1983[b], *Bifurcations of periodic solutions near equilibrium points of Hamiltonian systems*. In: Bifurcation theory and applications-Montecatini, 1983, 55-105, ed. L. Salvadori, LNM 1057, Springer Verlag, Berlin etc., 1984.

Dulac, M.H.: 1912, *Solution d'un système d'équations différentielles dans le voisinage de valeurs singulières*. Bull. Soc. Math. France 40, 324-383.

Eliasson, H.: 1984, *Hamiltonian systems with Poisson commuting integrals*. Thesis, University of Stockholm.

Euler, L.: 1764, *Considerationes de motu corporum coelestium* (E304). Novi Commentarii Academiae Scientiarum Petropolitanae 10. Also: Opera Omnia II, 25, 246-257.

Euler, L.: 1765, *De motu rectilineo trium corporum se mutuo attrahentium* (E327). Novi Comm. Ac. Soc. Petr. 11, 144- . Also: Opera Omnia II, 25, 281-289.

Euler, L.: 1772, *Theoria Motuum Lunae*. Typis Academiae Imperalis Scientiarum Petropoli. Also: Opera Omnia II, 22.

Giorgilli, A. and Galgani, L.: 1978, *Formal integrals for an autonomous Hamiltonian system near an equilibrium.* Cel. Mech. 17, 267-280.

Gustavson, F.G.: 1966, *On constructing formal integrals of a Hamiltonian system near an equilibrium point.* Astron. J. 71, 670-686.

Gyldén, H.: 1884, *Sur un cas particulier du problème des trois corps.* Bull. Astron. 1, 361-369.

Humphreys, J.E.: 1972, *Introduction to Lie algebras and representation theory*, 2nd printing. GTM 9, Springer Verlag, New York.

Izumiya, S.: 1982, *Note on stable equivariant maps.* Math. J. Okayama Univ. 24, 167-178.

Jacobi, C.G.J.: 1836, *Sur le mouvement d'un point et sur un cas particulier du problème des trois corps.* Compt. Rend. 3, 59 - .

Jacobson, N.: 1962, *Lie algebras.* Interscience tracts on pure and appl. math. 10, John Wiley and sons, New York.

Kelley, A.: 1963, *Using changes of variables to find invariant manifolds of systems of ordinary differential equations in a neighborhood of a critical point, periodic orbit, or periodic surface. History of the problem.* Technical report no. 3, Dept. of Math., Univ. of California.

Kummer, M.: 1981, *On the construction of the reduced phase space of a Hamiltonian system with symmetry.* Indiana Univ. Math. J. 30, 281-291.

Lagrange, J.: 1772, *Essai d'une nouvelle méthode pour resoudre le problème des trois corps.* In: Recueil des pieces qui ont remporté les prix de l'Academie Royale des Sciences (shortly Prix de l'Academie) IX, dated 1772 but not published until 1777, Panekoucke, Paris. Also Oeuvres, ed. M.J.A. Serret, vol. 6, Gautiers-Villars, Paris, 1873.

Lasalle, M.G.: 1973, *Une démonstration du théorème de division pour les fonctions differentiables.* Topology 12, 41-62.

Liapunov, A.: 1892, *Problème général de la stabilité du mouvement.* Ann. Math. Studies 17, Princeton Univ. Press, Princeton, 1947. (a reproduction of a french translation dated 1907 of a russian memoir dated 1892).

Martinet, J.: 1982, *Singularities of smooth functions and maps.* London Math. Soc. Lect. Note Series 58, Cambridge Univ. Press, Cambridge.

Mather, J.N.: 1968, *Stability of* C^∞*-mappings* I, *The division theorem.* Ann. of Math. 87, 89-104.

Mather, J.N.: 1969, *Stability of* C^∞*-mappings* II, *Infinitesimal stability implies stability.* Ann. of Math. 89, 254-291.

Mather, J.N.: 1968, *Stability of* C^∞*-mappings* III, *Finitely determined map germs.* Inst. Hautes Études Sci. Publ. Math. 35, 127-156.

Mather, J.N.: 1969, *Stability of* C^∞*-mappings* IV, *Classification of stable map germs by* $\mathbb{R}$*-algebras.* Inst. Hautes Études Sci. Publ. Math. 37, 223-248.

Mather, J.N.: 1970, *Stability of* C^∞*-mappings* V, *Transversality.* Adv. in Math. 4, 301-336.

Mather, J.N.: 1970, *Stability of* C^∞*-mappings* VI, *The nice dimensions.* In: Proc. of Liverpool Singularities Symposium I, ed. C.T.C. Wall, 207-248, LNM 192, Springer Verlag, New York.

Meer, J.C. van der: 1982, *Nonsemisimple 1 : 1 resonance at an equilibrium.* Cel. Mech. 27, 131-149.

Meyer, K.R.: 1974[a], *Generic bifurcations in Hamiltonian systems.* In: Dynamical systems-Warwick 1974, LNM 468, Springer Verlag, Berlin etc.

Meyer, K.R.: 1974[b], *Normal forms for Hamiltonian systems.* Cel. Mech. 9, 517-522.

Meyer, K.R. and Schmidt, D.: 1971, *Periodic orbits near* L_4 *for mass ratio's near the critical mass ratio of Routh.* Cel. Mech. 4, 99-109.

Moser, J.: 1958, *New aspects in the theory of stability of Hamiltonian systems*. Comm. Pure Appl. Math. 11, 81-114.

Moser, J.: 1968, *Lectures on Hamiltonian systems*. Mem. A.M.S. 81, 1-60.

Moser, J.: 1976, *Periodic orbits near an equilibrium and a theorem by Alan Weinstein*. Comm. Pure Appl. Math. 29, 272-747. addendum, ibid., 31, 529-530.

Pedersen, P.: 1934, *On the periodic orbits in the neighborhood of the triangular equilibrium points in the restricted problem of three bodies*. Monthly Not. Roy. Astron. Soc. 94, 167-185.

Poènaru, V.: 1976, *Singularités* C^∞ *en présence de symétrie*. LNM 510, Springer Verlag, Berlin etc.

Poincaré, H.: 1879, *Thèse*. Also Oevres I, 59-129, Gauthiers Villars, Paris, 1928.

Poincaré, H.: 1890, *La problème des trois corps et les equations de la dynamique*. Acta Math. 13, 1-270.

Poincaré, H.: 1892, *Les méthodes nouvelles de la mécanique céleste*. Dover publ. inc., New York, 1957. (Unaltered publication of the volumes originally published in 1892, 1893, 1899).

Pöschel, J.: 1982, *Integrability of Hamiltonian systems on Cantor sets*. Comm. Pure Appl. Math. 35, 653-695.

Poston, T. and Stewart, I.N.: 1978, *Catastrophe theory and its applications*. Pitman, London.

Roberts, M.: 1983, *Characterisations of finitely determined equivariant map germs*. Preprint I.H.E.S./M/83/57, Bures-sur-Yvette.

Robinson, R.C.: 1970, *Generic properties of conservative systems* I, II. Am. J. Math. 92, 562-603, 897-906.

Roels, J.: 1985, *Sur des nouvelles series pour le problème de masses critique de Routh dans le problème restreint plan des trois corps*. Cel. Mech. 12, 327-336.

Routh, E.J.: 1875, *On Laplace's three particles, with a supplement on the stability of steady motion.* Proc. London Math. Soc. 6, 86-97.

Schmidt, D. and Sweet, D.: 1973, *A unifying theory in determining periodic families for Hamiltonian systems at resonance.* J. Diff eq. 14, 597-609.

Schwarz, G.: 1975, *Smooth functions invariant under the action of a compact Lie group.* Topology 14, 63-68.

Siegel, C.L.: 1952, *Über die Normalform analytischer Differentialgleichungen in der Nähe einer Gleichgewichtslösung.* Nachr. Akad. Wis. Göttingen, Math. Phys. kl.IIa, 5, 21-36.

Siegel, C.L.: 1956, *Vorlesungen über Himmelsmechanik.* Springer Verlag, Berlin etc.

Smale, S.: 1970, *Topology and mechanics* I, II. Inv. Math. 10, 305-331; 11, 45-64.

Sternberg, S.: 1958, *On the structure of local homeomorphisms of euclidian n-space,* I. Amer. J. Math. 80, 623-631.

Sternberg, S.: 1959, *On the structure of local homeomorphisms of euclidian n-space,* II. Amer. J. Math. 81, 578-605.

Sternberg, S.: 1961, *Finite Lie groups and the formal aspects of dynamical systems.* J. Math. Mech. 10, 451-474.

Szebehely, V.: 1976, *Theory of orbits, the restricted problem of three bodies.* Academic press, London.

Takens, F.: 1974, *Singularities of vector fields.* Publ. I.H.E.S. 43, 47-100.

Vegter, G.: 1981, *The* C^p*-preparation theorem,* C^p*-unfoldings and applications.* Preprint ZW-8013, University of Groningen.

Volk, O.: 1976, *Miscellanea from the history of celestial mechanics.* Cel. Mech. 14, 365-382.

Volk, O.: 1983, *Eulers Beiträge zur Theorie der Bewegung der Himmelskörper*. In: Leonhard Euler 1707-1783, Beitrage zu Leben und Werk, Birkhäuser Verlag, Basel.

Wassermann, G.: 1977, *Classification of singularities with compact abelian symmetry*. Regensburger Math. Schriften 1.

Weinstein, A.: 1973, *Normal modes for non-linear Hamiltonian systems*. Inv. Math. 20, 47-57.

Weinstein, A.: 1978, *Bifurcations and Hamilton's principle*. Math. Zeitschrift 159, 235-248.

Whittaker, E.T.: 1902, *On the solution of dynamical problems in terms of trigonometric series*. Proc. London Math. Soc. 34, 206-221.

Whittaker, E.T.: 1917, *A treatise on the analytical dynamics of particles and rigid bodies*, 2nd ed. Cambridge Univ. Press, Cambridge.

Williamson, J.: 1936, *On the algebraic problem concerning the normal forms of linear dynamical systems*. Am. J. Math. 58, 141-163.

INDEX